Celebrating 50 Years 1964-2014

NATIONAL ACADEMY OF ENGINEERING

LETTER FROM THE PRESIDENT

Fifty years ago, on December 5, 1964, the National Academy of Engineering (NAE) was founded by the stroke of a pen when the National Academy of Sciences Council approved the NAE's articles of organization. The first NAE Council, made up of its 25 founding members, quickly elected Augustus B. Kinzel president and Eric Walker vice president. *The Making of the NAE: The First 25 Years* presented the history leading up to and spanning the Academy's founding. On the occasion of the Academy's 50th Anniversary the essays presented here highlight the prodigious changes in people's lives that have been created by engineering over the past half century and consider how the future will be similarly shaped.

In public discourse the words "engineering" and "science" are often used interchangeably but, as any scientist or engineer will confirm, they are entirely different pursuits. Science discovers and understands truths about the greater world, from the human genome to the expanding universe. Engineering, for its part, solves problems for people and society, ranging from such handy innovations as displaying room temperature on your cell phone to devising ways to protect buildings against earthquakes to the enormous complexity involved in putting a man on the Moon. Engineering solutions encompass airplanes and automobiles, information technology and communications, environment and health systems, sustainability and energy sufficiency, computers and space missions, and much more. All of these solutions have evolved over time as people and society required new technological capabilities to meet new needs. For instance, evolution in airplanes occurred when aluminum replaced wood in their construction and when jet engines replaced propeller-driven engines, so that travelers could fly faster, more comfortably, and more safely.

Over the past 50 years, engineering has transformed our lives literally every day and it will continue to do so going forward, utilizing new capabilities, creating new applications, and providing ever-expanding services to people. For example, not everyone thought they needed the iPhone when the first model was introduced in 2007. I recall the moment when I became hooked on mine. I was at my cabin in the mountains of California when, in an emergency, I urgently needed a veterinarian late on a Sunday afternoon. My iPhone directed me to the nearest open veterinary emergency room, which was 40 miles away in another state. Without the phone, I never would have found that vet.

My story—multiplied by the experiences of everyone who has adapted to new technology, whether the telephone and the automobile or the computer and the jet plane—captures the way engineering innovations quickly become integrated into our normal daily lives. These essays look at some of the areas of personal and societal life in which that seamless integration and the service provided has been evident over time. Three essays look at engineering innovations and solutions in the decades centered on 1964, 1989, and 2014. The fourth offers visions of what engineering may deliver in the next half century.

The fourth essay also reintroduces the NAE Grand Challenges for Engineering, a global vision proposed in 2008 describing urgently needed engineering solutions for tomorrow's engineers to tackle. Meeting the Grand Challenges—making solar energy economical, engineering better medicines, and providing access to clean water, among others—will make the planet not only "a more technologically advanced and connected place, but also a more sustainable, safe, healthy, and joyous place—in other words, a better place." The Grand Challenges represent the first global calling to engineering on behalf of the planet, a call that transcends countries, cultures, and continents.

The past half century verifies that by advancing science and engineering we gain an accelerating growth in both knowledge and technological capabilities that benefit people and society. I hope that these essays will underscore for you the depth, breadth, and importance of engineering to human health, wealth, and joy—to humanity itself.

— C. D. Mote, Jr.
President, National Academy of Engineering

1964

Dawn of the Digital Age

By the mid-1960s, thanks to the work of engineers in the decades just before and after World War II, Americans were accustomed to many conveniences in daily life. Tap water was safe to drink. Electric power was reliable and affordable. And the family could take its summer road trip on the new interstate highway system—including bridges, tunnels, rest stops for gas and food, and standardized signage—that connected an ever-growing number of cities and towns from coast to coast.

Today these conveniences are so commonplace that we think about them only when there's a problem—the power is out, there's a water main break, or two lanes on a bridge are closed for repairs. But when they first occurred, these advances and innovations had a profoundly transformative effect on the nation and on individuals and families. To cite just a couple of statistics: by the 1930s, the creation of sewage sanitation systems and public supplies of clean drinking water had virtually eliminated the spread of waterborne diseases like cholera and typhoid. Combined with other public health advances such as vaccination programs, antibiotics, and a safer food supply, those crucial improvements in sanitation and water supply helped increase the average life expectancy in the United States by 50 percent—from 47 years to 70 between 1900 and 1960. (By comparison, average life expectancy since the mid-1960s has increased only about 12 percent.) Similarly, by the 1940s, a few years after the establishment of the Rural Electrification Administration in 1935, 800 rural electric cooperatives had been formed and 350,000 miles of power lines brought millions of Americans in rural communities literally out of the "dark ages."

None of those advances happened by chance. In response to public demand, public policy, and an intrinsic creative drive, engineers created the infrastructure essential to the health, prosperity, and security of the American people—not only for electrification, sanitation, and water supply and distribution, but also for automobiles, highways, refrigeration, air-conditioning, aviation, high-performance materials, and much, much more. Nor did progress stop there. As would become clear in the latter part of the 20th century, engineering innovations between the mid-1940s and mid-1960s, many driven by Cold War national security concerns and the Department of Defense, were quietly laying the foundation for scores of advances that Americans in the 21st century would take for granted—advances in computers, communications, and health care, among other fields.

In 1964, glimmers of the changes that would transform American society were beginning to enter public awareness. Even as nuclear arms deployed by the United States and the Soviet Union in the ongoing Cold War loomed large, peaceful uses of atomic energy were emerging. The first nuclear power plants in the United States came online in the late 1950s; the use of nuclear medicine procedures for diagnostics and treatment, which had begun in the 1930s, expanded in the 1960s. Another recent invention, the laser, would soon demonstrate its value to health care and fiber-optic communications. And while the Air Force was using room-size mainframe computers to process data from far-flung radar stations and guard against attacks by Soviet bombers, the introduction of much smaller and more versatile computing machines was about to alter life in the United States and the world at large forever. A new era was dawning—a digital age that would transform how we lived, worked, and communicated.

ENIAC *(right)* ran on nearly 18,000 vacuum tubes and needed a staff to plug in thousands of wires to set or change its program. With the advent of the transistor *(far right)*, vacuum tubes became obsolete and computers began to shrink.

Electronics: Smaller, Faster, Cheaper

In the late 1940s and 1950s, electronic computers were still enormous and enormously expensive. They were the province of large institutions—governments, big corporations, universities, and especially the military—that could afford to buy them, build cooled rooms large enough to house them, and hire the operators to make them work. In 1946, a behemoth named ENIAC (Electronic Numerical Integrator and Computer) was unveiled at the University of Pennsylvania. It weighed 30 tons, occupied a room 30 feet by 50 feet, and operated with nearly 18,000 bulky, power-hungry vacuum tubes that frequently burned out. Commissioned to produce artillery firing tables so gunners in the field could adjust their aim as needed, ENIAC could perform in just 30 seconds calculations that used to take 12 hours on a hand calculator.

Powerful as it was, ENIAC had limitations. For one thing, this computer had only enough memory to handle the numbers involved in the current computation; its instructions, or program, had to be wired into the circuitry. So, changing the program meant someone had to spend several days unplugging and replugging thousands of wires to enter the changes and then test the new settings.

Even as ENIAC was coming online, engineers elsewhere were exploring a different way to build these machines. In 1947, a revolutionary engineering advance emerged from AT&T Bell Labs that would start society on the path to the current era of both hyper-fast supercomputers and the ubiquitous smartphone.

This innovation—a device based on solid-state semiconductor materials that could both amplify an electrical signal and turn it on and off—was the result of a brilliant collaboration among John Bardeen, William B. Shockley, and Walter H. Brattain. The team had been asked to develop a replacement for vacuum tubes, which were not only unreliable power hogs but also could not pick up the ultrahigh-frequency radio waves needed for AT&T's transcontinental telephone system. Two days before Christmas 1947, after a month of intense experimentation, the team presented their bosses at Bell Labs with the transistor *(above)*.

The transistor generated very little heat and was both dependable and tiny—characteristics that would lead to a phenomenal miniaturization of complex circuitry, paving the way for virtually every electronic device we rely on today. For their monumental "researches on semiconductors and their discovery of the transistor effect," Bardeen, Shockley, and Brattain shared the Nobel Prize in Physics in 1956.

Within a few years, engineers were using transistors to produce small devices that amplified sound, such as transistorized hearing aids and pocket-size transistor radios. By the

A SLICE OF LIFE

Way Back in the Pre-Digital Age

Although many large businesses, universities, and governments in the early 1960s used computers to keep track of payroll, print checks, and manage large databases for research, such machines were far from commonplace in ordinary households. A high school student in 1964 would prepare class assignments on a manual typewriter. Secretaries making multiple copies of documents typed with carbon paper and painstakingly corrected errors with a white "liquid paper" correcting fluid. The latest thing for taking snapshots was an "instant" camera that contained self-developing film and produced an image—after several minutes. If you wanted to send the photo to someone, you had to put it in an envelope, stick on a postage stamp, and mail it through the U.S. Postal Service. Telephones had rotary dials and no "call waiting." If the teenager in the house were on the line, a caller just had to try again later. "Leaving a message" meant getting through to a live person who had to write the message on a piece of paper and put it somewhere for the intended recipient. Oh, and when the phone rang? You actually had to answer it to find out who was calling.

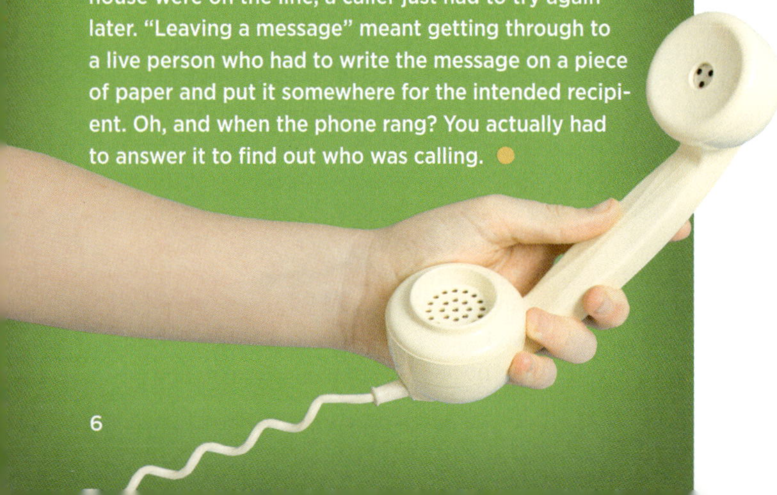

The elegant CDC 6600, with its plus-sign shaped panels, held the supercomputer speed record from 1964 to 1969.

mid-1960s, as transistor design and manufacturing improved, computer engineers used them to build a new generation of supercomputers, like Control Data Corporation's CDC 6600. Designed by Seymour Cray, the CDC 6600 was almost three times faster than the next fastest machine of its day, the IBM 7030 Stretch. Despite being phenomenally fast and much more reliable and efficient than ENIAC, the CDC 6600 was still a huge machine with a huge price tag. At $7 to $10 million apiece, it was not something your average business, and certainly not your average consumer, could afford.

The crucial engineering advance that brought computers out of large institutions and into much wider use was the integrated circuit, developed independently in the late 1950s by Jack Kilby of Texas Instruments and Robert Noyce of Fairchild Semiconductor, a pioneering firm in California's Silicon Valley. In 1989, Kilby and Noyce would be awarded the first Charles Stark Draper Prize for Engineering, the National Academy of Engineering's (NAE's) highest award, for "their independent co-invention of the monolithic (meaning formed from a single crystal) integrated circuit, better known as the semiconductor microchip." Robert Noyce died in 1990, but in 2000 Jack Kilby was awarded half of that year's Nobel Prize in Physics "for his part in the invention of the integrated circuit." (The other half of the prize was shared by Zhores I. Alferov and Herbert Kroemer "for developing semiconductor heterostructures for high-speed- and opto-electronics.")

The integrated circuit squeezed multiple transistors, wiring, and other components of an electronic circuit onto a single silicon chip using photographic techniques to reduce the circuit design to a tiny imprint, which was then printed on a wafer the size of a baby's fingernail. Integrated circuits produced in the 1960s were essential to early aerospace projects such as the Minuteman missile and the Apollo program, which both needed lightweight digital computers for their inertial guidance systems. This early government support allowed integrated-circuit makers to refine manufacturing methods and lower costs enough to enter the industrial and, eventually, the consumer markets.

As production costs came down, the average price per integrated circuit dropped from $50 in 1962 to $2.33 in 1968, even as the number of transistors on a chip skyrocketed. In 1965 Gordon Moore—who worked with Noyce at Fairchild Semiconductor and later joined him as cofounder of Intel Corporation—predicted that computing capacity, based on the number of transistors packed into a chip, would double every year. The race toward ever smaller yet ever more powerful computers was off and running. (Updating his forecast in 1975, Moore predicted that chip capacity would double every two years, an estimate that remained close to the mark for decades. "Moore's Law" is still used today as a standard for measuring industry progress—a testament to the creativity and ingenuity of engineers focused on improving both performance and cost.)

Computers as we know them would not exist, of course, without the ingenuity of the programmers and software engineers who created the programming languages, operating systems, and applications that make the machines useful in so many different ways. High-level programming languages like Fortran, COBOL, and BASIC were instrumental in making programming faster and considerably less tedious than hand-coding in the ones and zeros of machine language.

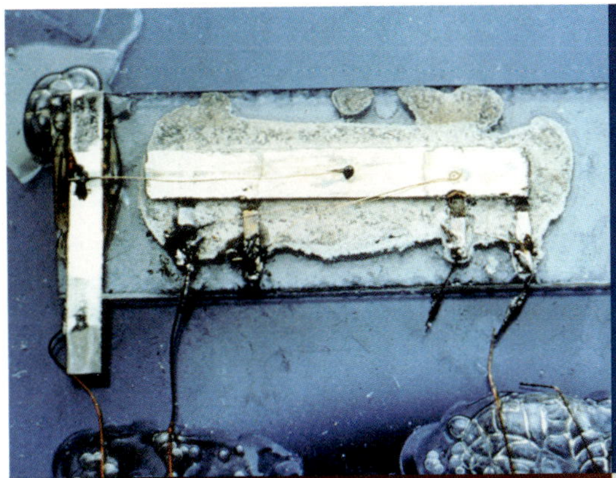

From a few transistors in the first integrated circuit *(left)*, the number of components crammed on a microchip doubled every two years, as predicted by Gordon Moore (shown seated, *below*, with Robert Noyce, one of the inventors of the integrated circuit). In 1995, University of Pennsylvania engineering students designed "ENIAC on a chip"—recreating the 30-ton ENIAC's circuits with 250,000 transistors on a chip only 8 mm square *(below, left)*.

FORTRAN

(FORmula TRANslating System) was developed in the mid-1950s by an IBM team led by John Backus. "Much of my work has come from being lazy," Backus told *Think,* the IBM employee magazine, in 1979. "I didn't like writing programs, and so, when I was working on the IBM 701 [an early computer], writing programs for computing missile trajectories, I started work on a programming system to make it easier to write programs." Designed for scientific and engineering applications, some version of Fortran is still used in intensive supercomputing tasks such as weather and climate modeling, computational fluid dynamics, and structural engineering. John Backus was awarded the NAE's Draper Prize in 1993 for "development of FORTRAN, the first widely used, general purpose, high-level computer language."

COBOL

(COmmon Business-Oriented Language) was created by a committee of computer manufacturers and their clients, notably the government. A key member of the committee was the indomitable programmer Rear Admiral Grace Murray Hopper, who had long believed that programming languages should be usable by people who were neither mathematicians nor computer experts. The goal was to create a language suited to large-scale data processing such as for payrolls, budgets, and inventory—and to have programs that could run on different makes of machines. This compatibility was especially important to the Department of Defense (DOD), which bought computers from different manufacturers. In December 1960, the same COBOL program ran successfully on both a Remington Rand UNIVAC II and an RCA 501. COBOL would dominate government and business data processing for decades and is still used for millions of banking transactions today.

BASIC

(Beginner's All-purpose Symbolic Instruction Code) was invented in 1963 at Dartmouth College by mathematicians John Kemeny and Thomas Kurtz *(below)* as a teaching tool for undergraduates. Kemeny and Kurtz had the radical idea that undergrads—science and nonscience majors alike—could learn about computing by actually writing their own programs. But first they needed a more user-friendly language. The language they created used simple English words such as PRINT, SAVE, and RUN. To get the computer to write something you merely typed PRINT, followed by the words to print in quotes. Kemeny wanted the language to be so easy that a complete novice "could use it after three hours of training." Versions of BASIC became popular with the advent of minicomputers such as Digital Equipment Corporation's PDP line in the mid-1960s and then exploded with the introduction of home computers in the mid-1970s. (Bill Gates and Paul Allen wrote a version of BASIC for the MITS [Micro Instrumentation Telemetry System] Altair and then went on to form Microsoft—and the rest, as they say, is history.)

THE MINICOMPUTER

Affordable, Compact, and User-Friendly

In 1965, Digital Equipment Company introduced the PDP-8—eighth in a revolutionary line of interactive computers that focused on the user's experience rather than solely on machine efficiency. Sold for $18,000 and available in a desktop configuration, the PDP-8 was the first commercially successful minicomputer, affordable for many midsize businesses and small laboratories. A Digital executive in England, where small cars and short skirts were in fashion in the 1960s, was credited with coining that term in a sales report: "Here is the latest minicomputer activity in the land of miniskirts as I drive around in my [Austin] Mini Minor." To promote the machine's small size, the company photographed it in the back of a Volkswagen Beetle *(above)*. Soon the PDP-8 was at work in many settings, from controlling the baseball scoreboard at Boston's Fenway Park *(opposite, top)* and the lights at a New York theater to doing signal analysis in physics labs and monitoring instruments in a hospital operating room *(opposite, bottom)*. In 1970, the PDP-8/E came along, priced at only $6,500, with a configuration that allowed devices such as teletypewriters and line

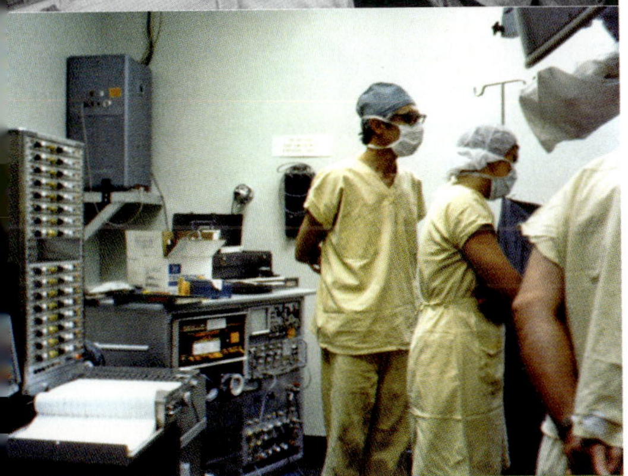

printers to be connected to it. The PDP-8's influential successor, the PDP-11, would become the go-to hub for computer "time-sharing" at universities and the platform for developing the widely used C programming language and the UNIX operating system.

PDP stood for "programmed data processor," a term chosen to avoid the stereotype that "computers" were too big, too expensive, and required a big staff. The PDP line's interactivity inspired programmers to create not only early text-editing and music programs but also games, including Spacewar!—the first computer video game.

Harnessing Light: The Power of Lasers and Fiber Optics

The availability of electricity lit the world through incandescent light bulbs in the early 20th century, but it wasn't until the 1960s and later that the marvels of intense laser light and fiber optics were realized. Laser light depends on the phenomenon of stimulated emission of radiation, theorized by Albert Einstein in 1917. Several decades would pass before the engineering expertise of Charles Townes, Arthur Schawlow, and Theodore Maiman turned the theory of stimulated emission into a creation that would serve society.

Townes was both a physicist and a skilled engineer. During World War II, he had worked as an engineer at Bell Labs, developing guidance systems using microwaves. He later credited "my experience in both engineering and physics" as crucial to the advances he made. In 1954, Townes and colleagues at Columbia University built the first maser (microwave amplification by stimulated emission of radiation). In 1958, Townes and Schawlow of Bell Labs (who also happened to be Townes's brother-in-law) theorized that masers could be made to work at optical and infrared wavelengths and proposed a way to generate a strong beam by amplifying the stimulated emission in a cavity bounded with mirrors. In May 1960, Maiman, of Hughes Research Laboratories, a division of Hughes Aircraft Company, used that mirror technique to produce the first laser (light amplification by the stimulated emission of radiation), by energizing chromium atoms in ruby crystals.

Many skeptical scientists in 1960 considered the laser "a solution looking for a problem," Townes noted. "But by bringing together optics and electronics, lasers opened up vast new fields of science and technology." The ruby laser was only the first of a plethora of types that engineers went on to devise. In the decades since, more than 50,000 engineering patents have been issued for devices and techniques involving lasers. In 1964, Townes was awarded a Nobel Prize for his role in developing the maser and also for conceptualizing its super-adaptable successor, the laser.

Lasers proved useful to the nation's security as range finders for guns and as target designators for guided weapons. They became equally useful in a host of nonmilitary applications, from reading digital data on barcodes and digitized music on CDs to increasing the capacity of landlines to handle both phone calls and the growing digital traffic of computer networks. Beginning in the 1970s, amazing breakthroughs in the manufacture of low-loss optical fiber by AT&T and Corning would make possible the fiber-optic cables that form the backbone of today's information grid.

In health care, lasers greatly advanced methods of treating disorders of the eye. As early as November 1961, 18 months after the demonstration of the device, physicians used a ruby laser to treat a retinal tumor. In 1964, William Bridges of Hughes Research Labs devised an argon laser to reattach detached retinas, a condition which, if left untreated, can result in blindness. This was a major improvement over cauterization with extreme heat, the original treatment for a detached retina. Laser operations would eventually save the sight of millions of people with diabetic retinopathy and correct the vision of millions more.

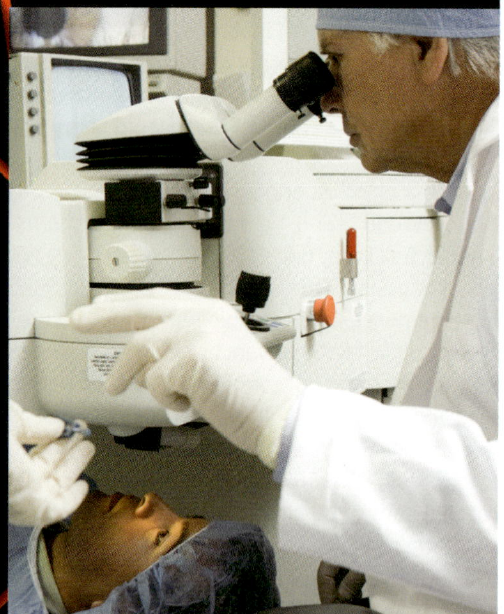

The Multi-Talented Laser

Essential instrument for skin and eye surgery, precision machining tool, booster for fiber optic communications—the laser is all this and more. But it is most familiar as the barcode scanner at grocery and other retail stores where it revolutionized the checkout line.

The World Above and Beyond: Space Exploration

Perhaps no engineering achievement of the 20th century was more inspiring to the American people than the moment that *Apollo 11* astronaut Neil Armstrong became the first human to set foot on the moon on July 20, 1969. Everyone old enough to have been awake and aware that day can remember where they were when they saw the event on television or heard it on the radio. It was both a proud step for the nation whose flag Armstrong planted on the moon and a "giant leap for mankind," as he put it. Getting the crew of *Apollo 11* to the moon's surface and back home to Earth was an incredible engineering undertaking.

When the Soviets launched the first satellite, Sputnik I, in October 1957—beating the first U.S. satellite by several months—Americans were shocked. At the height of the Cold War, it was clear that rockets powerful enough to lift satellites into outer space could also target distant cities with nuclear-armed intercontinental ballistic missiles (ICBMs). Sputnik alerted Americans that their country did not lead in every area of science and engineering and needed to reassess its policies and priorities. The realization led to a significant boost in federal support for scientific education and technological research.

In May 1961, a month after Russian cosmonaut Yuri Gagarin became the first man in space—chalking up another "first" for the Soviets—President John Kennedy committed the United States to "achieving the goal, before this decade is out, of landing a man on the Moon and returning him safely to the Earth." Although one objective of that program was to assert American engineering superiority over the Soviets, reaching for the moon inspired humanity with a quest that transcended the Cold War.

A photograph of Earth, taken from lunar orbit on Christmas Eve 1968, gave humanity a new perspective on our planet. A few months later, Neil Armstrong took a "giant leap for mankind"—and then photographed Buzz Aldrin *(far right)* stepping onto the Moon's surface after him.

Designing the spacecraft and methods of propulsion, communication, and life support needed to achieve this audacious goal ranks as one of the great systems engineering achievements of all time. Everything from developing materials for the heat shield to withstand temperatures greater than 5000°F during reentry into Earth's atmosphere to designing the astronauts' space suits was an enormous engineering challenge. The mission also needed a computerized inertial guidance system to determine how much rocket thrust to apply for critical maneuvers such as landing the lunar module on the moon, linking it again with the service and command modules for the return trip, and achieving the proper reentry angle to Earth's atmosphere.

Professor Charles Stark Draper of Massachusetts Institute of Technology (MIT) took charge of developing the guidance system. With the myriad challenges involved in making sure that gyroscopes and accelerometers could function reliably and accurately in space, Draper knew his job would be a race against time, but he promised National Aeronautics and Space Administration (NASA) administrators that the system would be ready "before you need it." One very big piece of good luck: the miniaturization of computers had progressed sufficiently to allow Draper's team to produce equipment small enough to install in modules where space was tight. Without such computerized guidance, President Kennedy's goal would have been unattainable.

With each launch of an Apollo mission, the space program ignited the public imagination. On Christmas Eve 1968, seven months before the moon landing, *Apollo 8* astronaut William Anders took the famous *Earthrise* image that would become a moving reminder that Earth was the precious inheritance and responsibility of all humanity, "to be handled with utmost care," as Anders put it.

The lunar landings ended in 1972, but other space-related activities would continue, producing such long-term benefits for society as advances in robotics, solar power, and biomedical research. Perhaps the most familiar space-based benefits from the 1960s and '70s are the multitude of satellites that bring us television broadcasts, up-to-the-minute weather data, and pinpoint navigation.

Two American engineers—John Pierce of Bell Labs and Harold Rosen of Hughes Aircraft Company—developed key technologies in the 1950s and '60s that made commercial communication satellites possible. Pierce calculated the precise power needed to transmit signals to satellites in various Earth orbits and devised something called a traveling wave tube amplifier, which enabled a satellite to receive, amplify, and transmit radio signals. Rosen engineered spin-stabilization technology to aim the satellite's antennas for both receiving and transmitting signals. In 1995 the two shared the NAE's Draper Prize "for development of communication satellite technology." In October 1964, Syncom 3, the first geostationary communications satellite, relayed live television broadcasts of the Tokyo Olympics. The following year NASA turned the satellite over to DOD, which used it for, among other things, communicating with commanders in Vietnam.

Weather forecasters gained invaluable, lifesaving tools with the launch of TIROS-1, the first weather satellite, on April 1, 1960, followed by the ESSA (Environmental Science Services Administration) satellites in February 1966, which provided cloud-formation photography to the Weather Bureau's National Meteorological Center. This global weather satellite system transmitted thousands of images back to Earth, enabling ground station forecasts of weather patterns, including hurricanes.

In 1973, a multiservice Joint Program Office within DOD began developing a satellite-based radio navigation system for delivering weapons precisely on target. The system became operational for the military in the mid-1980s, with a coarser version accessible to the public. In 2000, President Bill Clinton made access to more precise signals fully available to the general public. Today, along with similar satellite systems launched by other nations, the global positioning system, or GPS, helps guide civilians around the world to their destinations through a variety of GPS devices.

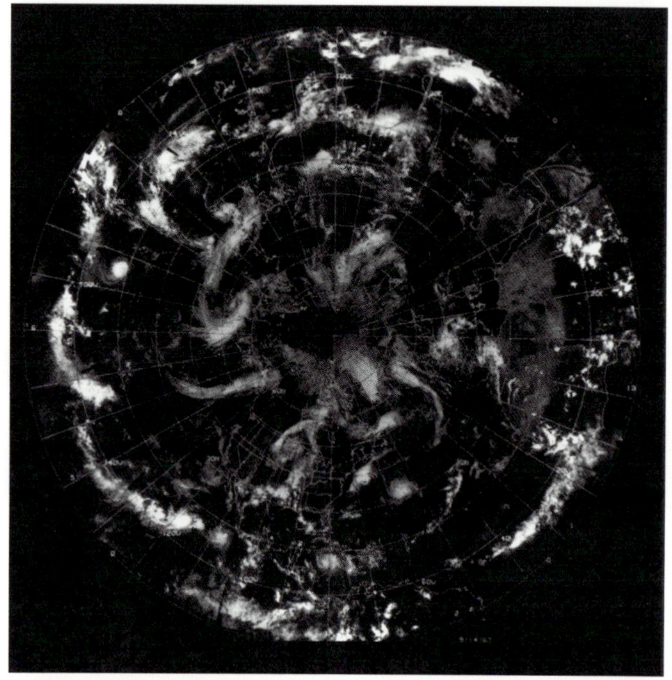

With the launch of observational satellites in the 1960s (mosaic of ESSA-5 images, *top*), meteorologists were able to view weather systems over large areas of the planet and spot storms in the making (NIMBUS image of Hurricane Alma in August 1966, *bottom*).

The versatile and reliable Boeing 727 was the best-selling airliner in the world during the first 30 years of the jet age.

Jet flights had a broad impact on American society, facilitating travel for tourism and business not only nationally but internationally as well.

Up, Up, and Away

While astronauts were blasting off for the moon in the 1960s, millions of people on Earth also began soaring to new heights of their own. Carried by commercial jets that cruised at altitudes far above those of propeller-driven planes, air travelers could avoid storms and enjoy safer, more comfortable flights.

Hans von Ohain in Germany *(right)* and Frank Whittle in England *(far right)* independently developed the jet engine, which revolutionized aviation.

Two engineers who had been on opposite sides during World War II developed the jet engine independently and almost simultaneously. Frank Whittle in England and Hans J. P. von Ohain in Germany later became good friends (in 1991, they were jointly awarded the Draper Prize for their work). Although not ready in time to affect the war, the turbojet engine revolutionized aviation and the postwar world.

The new aircraft were a testament to the work of aeronautical engineers, who had to design wings sturdy enough to endure speeds exceeding 500 miles per hour and airframes strong enough to sustain the material fatigue caused by vibrations and many cycles of pressurizing (required for those not wearing oxygen masks) and depressurizing cabins. To ensure safety and avoid costly modifications after planes entered production, engineers needed a reliable method for determining in advance whether their designs could withstand the stresses of flight. M. Jon Turner, head of Boeing's Structural Dynamics Unit, addressed that problem in the early 1950s by bringing civil engineering professors Ray Clough of the University of California, Berkeley, and Harold Martin of the University of Washington to Boeing for summer "faculty internships." Collectively, they created a method of structural analysis that Turner applied at Boeing using computers to perform the myriad calculations needed to predict real-world performance.

That fruitful collaboration led to Clough's development a few years later of what he named the finite element method (FEM). Clough formed a research group at UC Berkeley that used FEM in a host of analytical and experimental activities, from designing buildings and structures to withstand nuclear blasts or earthquakes to analyzing structural requirements for spacecraft and deep-water offshore drilling. By revolutionizing the application of computer technologies in engineering, FEM con-

tinues to help engineers design to this day all sorts of durable, cost-effective structures.

Meanwhile, Turner's efforts at Boeing contributed to the success of its renowned line of commercial jets, beginning in 1958 with the 707 and continuing in 1964 with the 727, which could land on shorter runways and serve more airports. Equipped with three fuel-efficient turbofan engines, the 727 became the workhorse of commercial aviation and helped achieve a threefold increase in U.S. passenger air traffic in the '60s.

Jet flights had a broad impact on American society, facilitating travel for tourism and business not only nationally but internationally as well. As the cost of air travel came down, more people took to the skies and the flight paths of aircraft crisscrossing the globe seemed to pull the continents closer together. By the end of the 1970s, an experience that had once been out of reach for most ordinary Americans—so exclusive, in fact, that air travelers actually dressed up for their flights *(above)*—became the way college students in jeans and sneakers would go home for Thanksgiving.

The World at Home: The Environmental Challenge

The American environmental movement—born in the late 1800s when naturalists like John Muir campaigned to protect wilderness areas—broadened in the mid-1900s as environmentalists drew attention to the far-reaching impact of pollution from a variety of sources in the developing country. Much of the environmental damage stemmed from the unforeseen consequences of solutions to earlier challenges. The pesticide DDT, for example, was so good at killing the insects that ravaged crops or transmitted diseases like malaria that the chemist who refined it in 1940, Paul Müller of Switzerland, won a Nobel Prize. Not until 1962, when biologist Rachel Carson's influential book *Silent Spring* appeared, did public attention in the United States focus on the hazardous side of DDT and other powerful pesticides. Sprayed over wide areas, these pesticides killed not only the targeted pests but many beneficial insects as well. DDT also entered the food chain of birds and other animals, some of which were threatened with extinction as a result. In humans, a growing body of evidence linked DDT to breast cancer, diabetes, and impaired neurodevelopment in children.

After a decade of considerable controversy, DDT was banned in the United States in 1972. Its use would also be discontinued in much of the rest of the world, although, in the absence of an equally effective and inexpensive chemical substitute, DDT would remain in limited use in countries where malaria is endemic.

On December 17, 1963, President Lyndon Johnson signed into law the Clean Air Act of 1963, which set emissions standards for power plants, steel mills, and other stationary sources, and recommended emissions standards for vehicles, which would be established by law in 1965. Over the next decade, with support from both major parties, Congress placed further limits on air and water pollution. Engineering met the challenge of new emission standards by developing new instruments to measure or reduce pollutants and new methods to upgrade or replace inadequate technologies.

Sometimes the dangers of a particular technological solution were suspected but mostly ignored until research findings created societal pressure for a different solution. Tetraethyl lead, for example, was added to gasoline starting in the 1920s, to prevent a phenomenon in auto engines called knocking—sudden bursts of combustion that can damage engines and reduce fuel efficiency. Although lead poisoning has been known since antiquity, and although

The June 1969 fire on Cleveland's Cuyahoga river *(top)*, helped spur the environmental movement and the Clean Water Act. On the fire's 40th anniversary *(above)* the Cuyahoga was sparkling.

In 1970, the first Earth Day was observed and the Environmental Protection Agency (EPA) was established.

manufacturers of tetraethyl lead had learned in the 1920s that without strict controls in factories workers would go insane and die of lead poisoning, it was not until the mid-1960s that more precise lab techniques could calculate the impact of lead exposure on human beings.

Credit for engineering those techniques belongs to geochemist Clair Patterson, who in 1965 warned that leaded gasoline and other industrial products were exposing people to far greater concentrations of lead in air and water than existed prehistorically. He found that concentrations of lead in modern human tissue were many times greater than in ancient human bones. Lead is now known to be hazardous in concentrations as low as 0.15 microgram per cubic meter of air, equivalent to less than one part per billion. Toxic to human organs and tissues, lead also interferes with a variety of physiological processes, including development of the nervous system, which means that children exposed to lead can suffer permanent learning and behavior disorders.

Patterson's findings led to the virtual elimination of lead in gasoline in the mid-1970s because society declared that the cost to human health and the environment vastly outweighed the benefit provided by leaded gasoline. The ban did not result in a loss of automotive performance, however, thanks to engineering innovations such as redesigned engine valves and safer additives for gasoline. (A federal ban on lead-based paint followed in 1978.)

Another example of environmental gains achieved because of public pressure—and the ingenuity of engineers in response to society's demands—began on June 22, 1969, when a big oil slick in Cleveland's notoriously polluted Cuyahoga River caught fire and damaged two bridges before firefighters extinguished it. Fires on the Cuyahoga were common, but in this instance *TIME* magazine ran a photo of the Cuyahoga in flames to illustrate the plight of the nation's waterways, which it dubbed "America's Sewage System." The Cuyahoga fire alarmed the public and boosted support for the environmental movement. In 1970, the first Earth Day was observed and the Environmental Protection Agency (EPA) was established. Two years later Congress passed the Clean Water Act, which protected rivers, estuaries, bays, and wetlands by regulating the discharge of pollutants within the nation's watersheds.

The environmental movement and the laws arising from it would provide impetus to formation of a new discipline, environmental engineering, which had long been a concern of the civil engineers who developed water and sewage systems essential to public health. Environmental engineering emerged as a distinct academic and professional field in the late 1960s and '70s as the need arose for creating solutions to environmental problems involving infinitesimal amounts of pollutants.

To Your Health: The Engineering of Medical Imaging and Therapies

The 1960s saw a flowering of medical engineering advances that built on work during and just after World War II. Nuclear medicine imaging, for example, uses radioisotope tracers developed at the beginning of the war at MIT, Brookhaven National Laboratory, and the University of California's Radiation Laboratory in Berkeley (later Lawrence Berkeley Laboratory). Inserted into the bloodstream, the radiotracers accumulate in areas of high chemical or metabolic activity, where they emit a small amount of radiation that can be detected to reveal tumors and other disorders. In the 1950s, two methods were developed for detecting this radiation for medical diagnosis—the Anger camera, developed by electrical engineer Hal Anger at Berkeley's Radiation Laboratory, and positron emission tomography (PET), developed by Gordon Brownell, head of the Physics Research Laboratory at Massachusetts General Hospital (MGH), and William Sweet, Chief of the Neurosurgical Service at MGH.

Medical ultrasonography, which doesn't require the use of radiotracers, had its origins in the wartime technology known as SONAR (SOund Navigation And Ranging). It involves sending out pulses of sound and recording "echoes" to produce images that allow doctors to detect tumors, lesions, and other abnormalities in the heart and other organs as well as in tendons, muscles, and blood vessels. By the mid-1960s, ultrasound was becoming a familiar tool in obstetrics to check the health of a fetus in the womb.

Other now-familiar medical engineering tools came along in the 1970s. Magnetic resonance imaging, or MRI, uses a magnetic field and radio waves to create detailed images to help diagnose a variety of problems, including aneurysms, disorders of the eye, damage from heart attack or heart disease, and joint disorders from arthritis. Also in that decade engineer Godfrey Hounsfield, of Britain's EMI Laboratory, and South African–born American engineer Allan Cormack of Tufts University independently devised 3-D imaging methods known as X-ray computer-assisted tomography (CAT). CAT scans would become the primary tool for diagnosing brain and spinal disorders.

These mid-century advances in medical imaging were in good part the product of digital engineering that allowed the translation of voltage signals into words and images displayed on computer monitors. By the 1980s many doctors used such computerized scans to reassure patients or diagnose ailments promptly, without the need for invasive diagnostic surgery. By enabling the early detection and treatment of many types of cancer as well as

Hal Anger, of the Radiation Laboratory, developed the Anger scintillation camera to detect radiation for medical diagnosis.

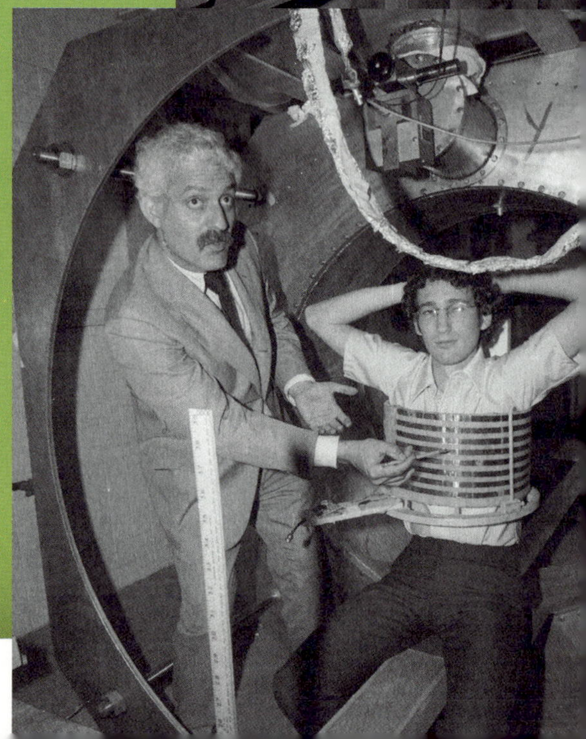

Dr. Raymond Damadian *(below, standing)*, one of the inventors of MRI, and Dr. Laurence Minkoff demonstrate a "super magnet" that would soon be used to gain information about the interior of the body without surgery.

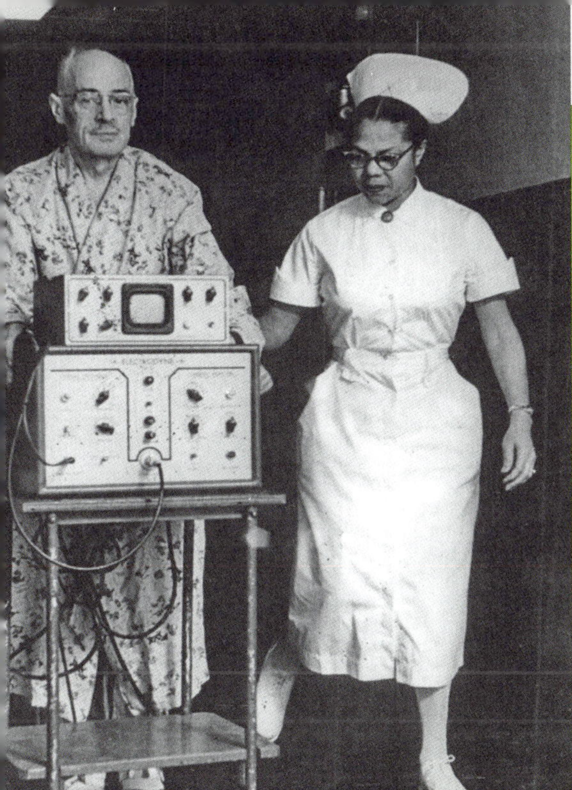

Early pacemakers were bulky boxes of electronics that kept a patient tethered to the nearest electrical outlet *(above)*. In the 1960s, an implantable pacemaker developed by Wilson Greatbatch of the University of Buffalo *(left)* returned heart patients' freedom of movement.

brain disorders, heart and vascular problems, and other diseases, these scans continue to save countless lives.

Sometimes an engineering solution is needed to regulate, repair, or replace an ailing heart or other organ. In 1957, electrical engineer Earl Bakken developed the first wearable pacemaker, a device that regulates the heartbeat by applying imperceptibly small electric impulses to heart muscles. Bakken's battery-powered, handheld pacemaker allowed patients in hospitals to move around. The first long-lasting implantable pacemaker was invented a couple of years later by electrical engineer Wilson Greatbatch, who miniaturized his device using silicon transistors. Greatbatch teamed up with two surgeons, who experimented successfully on animals before implanting one of his pacemakers in 1960 in a critically ill heart patient; the patient lived for another 18 months with the device. In the early 1970s, Greatbatch replaced the mercury battery in his pacemakers with a durable lithium battery that could last 10 years or more, reducing or eliminating the need for frequent operations to replace the battery. In 2001 Bakken and Greatbatch shared the NAE's inaugural Fritz J. and Dolores H. Russ Prize "for independent development of the implantable cardiac pacemaker."

People with kidney or heart disease are indebted to a gifted doctor who was also an exceptional engineer. Dr. Willem Kolff, a Dutch-born physician, developed the first kidney dialysis machine during World War II before moving to the United States to work on artificial organs. Beginning in 1967 he led a long-term effort by doctors, scientists, and engineers at the University of Utah to produce the first permanent artificial heart. In 1982 an artificial heart designed by Dr. Robert Jarvik in conjunction with Kolff and other team members was implanted in a patient near death, who survived for 112 days. With continued engineering improvements, artificial hearts, prolonged patients' lives for a few years until they could receive natural heart transplants. Kolff was awarded the 2003 Russ Prize for "pioneering work on artificial organs, beginning with the kidney, thus launching a new field that is benefitting the lives of millions."

Starting in the late 1960s, the combination of engineering and medicine became a potent academic program for creating innovations that can save life, extend life, or improve life. Biomedical engineering programs are now widespread, with women making up nearly 40 percent of those earning degrees in that field, the highest percentage in any engineering field other than environmental engineering.

The impact of bioengineers has been enormous. They play a key role in making many of today's breakthrough drugs practicable, for example, and advances in medical imaging and implantable medical devices have dramatically changed both diagnosis and treatment for people who become injured or ill.

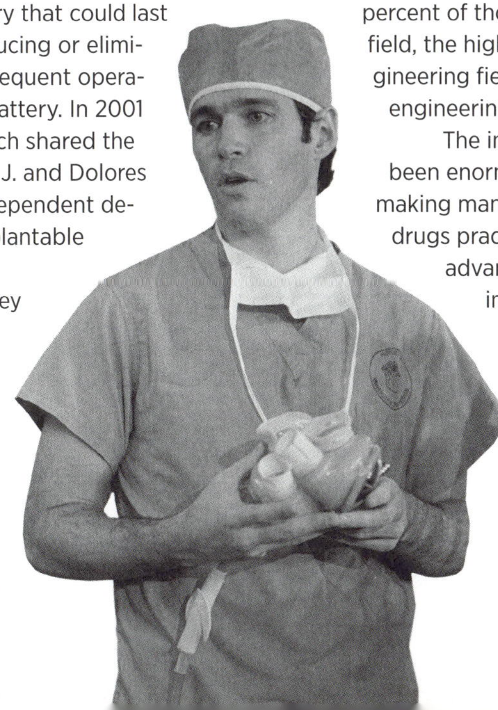

Dr. Robert Jarvik and the first permanent artificial heart.

1989

The Shrinking Globe

The phenomenon known as globalization—the growing interconnectedness of the world's peoples, economies, and cultures—has been accelerating since the 1970s, spurred by engineering advances in transportation, production, communication, and, most of all, in computer and information technology. By 1989, as the National Academy of Engineering celebrated its 25th anniversary, more than a dozen countries had connected to the fast-growing Internet, and the birth of the World Wide Web was just around the corner.

When the Berlin Wall came down in November 1989, signaling an end to the Cold War, the lowering of travel, political, and economic barriers between East and West raised the likelihood of more international cooperation on matters of both planet-wide and national concern. Just the month before, the National Institutes of Health (NIH) in the United States had launched the Human Genome Project. Within a few years the project would grow into an international collaborative effort to decipher the human genetic code governing heredity and its consequences, including hereditary diseases.

As the world shifted to the idea of international partnerships and engagement, engineers grappled with global issues such as improving energy efficiency and coping with climate change. It was becoming increasingly clear that no country—not even one as powerful as the United States—could remain secure and prosperous or solve far-reaching global problems, such as air pollution, on its own. "Just as pollutants flow from nation to nation, so capital and technological knowledge flow across national borders," remarked Robert White, NAE president, at its 25th annual meeting. "In short, our national interests can be served only by global bargains of interdependent nations—economic, industrial, and environmental bargains."

J. C. R. ("Lick") Licklider envisioned a network of connected computers that gave users access to programs and data anywhere. In 1969, that ancestor to the Internet consisted of just four nodes *(below)*, but was poised for rapid growth.

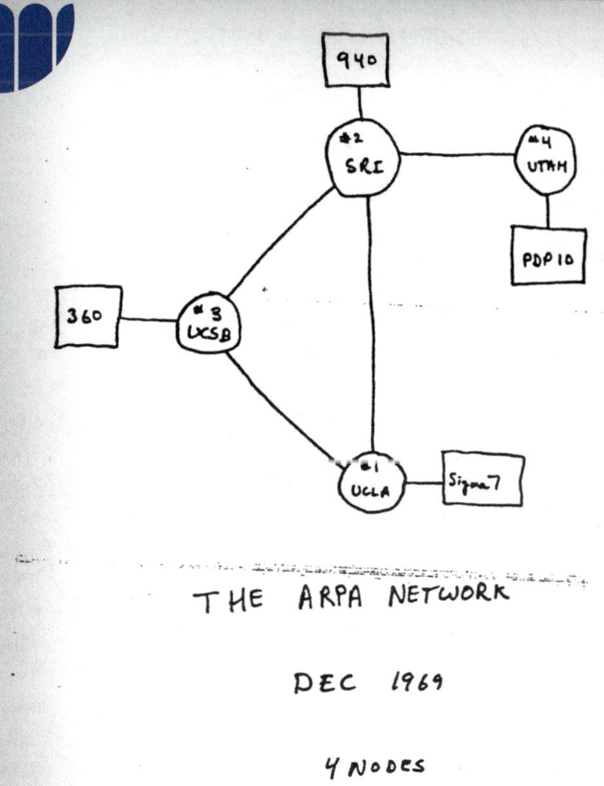

Building the Digital Highway

Interdependence, connection, and collaboration were values that drove the engineers who created the Internet. J. C. R. Licklider, the visionary first director of the Information Processing Techniques Office (IPTO) at DOD's Advanced Research Projects Agency (ARPA) in the early 1960s, was chief evangelist for a radical new idea: connecting computers in such a way that all users could have access to data and software from anywhere. During his tenure at IPTO, Licklider funded research for three seminal developments in information technology—creation of computer science and engineering departments at several major universities, time-sharing, and networking. His ideas and the work of the many people he sponsored led, directly or indirectly, to the interconnected information age we live in today.

Licklider called his idea the "Intergalactic Computer Network"—a tongue-in-cheek recognition of how far-fetched a widespread computer network seemed at the time. "We didn't really expect to get at that right away," he remembered later. "It was all we could possibly do to make time-sharing systems work." Time-sharing—letting multiple people connect to the same mainframe and use its power seemingly simultaneously—was still in its infancy. This was, after all, the era of "batch processing," which often meant waiting in line (sometimes for days) for your job to be run on a big mainframe. Creating a network connecting two or more mainframes with time-sharing capabilities was radical science fiction.

It took a decade, and the efforts of numerous teams of engineers working on both hardware and software—including pioneering work on packet-switching by Leonard Kleinrock at UCLA and Lawrence Roberts at ARPA—but in October 1972, science fiction became reality. ARPANET made its debut at the International Conference on Computer Communication (ICCC) in Washington, DC, and demonstrated the viability of its new packet-switching network technology by connecting 20 computer "nodes" located at universities and other sites around the country. Engineers working on advancing what would become the backbone of the information age suddenly had a new and powerful tool that could make it happen.

Soon a number of commercial enterprises as well as various academic computer research centers were developing computer networks of their own. Each of these networks had its own set of largely incompatible languages, operating systems, and protocols, so former ARPANET designers Robert Kahn and Vinton Cerf began laying the foundation for the Internet-to-come by developing two sets of protocols (usually referred to jointly as TCP/IP, for Transmission

Control Protocol/Internet Protocol) that allowed computers and networks to communicate with one another regardless of what software or hardware they used. The burgeoning network led, in the early 1980s, to the Domain Name System (DNS), a kind of automatically updated phone book of host computers and their numerical addresses, which created the original domains of .org, .gov, .com, .edu, .net, .mil, and .us.

In 1985, the National Science Foundation (NSF) announced the creation of five supercomputing centers to meet the U.S. research community's growing need for access to massively high-speed computers. A key part of this initiative was creation of NSFNET, which the NSF envisioned as a general high-speed network connecting the supercomputing centers to regional networks, local academic networks, and ARPANET—creating, in other words, a "network of networks" or "inter-net." Moreover, NSF decided to make NSFNET available not just to users at supercomputing center but to all academic users. Before the end of the decade, the Internet would go international.

Traffic on the Internet grew so quickly that NSF soon realized it needed more capacity. At the same time, NSF sought the participation of the private sector, opening the digital highway to commercial traffic in order to support networking, build volume, and bring costs down for everyone. Of course, with every upgrade to the infrastructure backbone, there was a surge in demand. By 1992, more than 100 countries and more than 6,000 networks were connected—and one-third of those networks were located outside the United States. In 2001, Cerf, Kahn, Kleinrock, and Roberts were awarded that year's NAE Draper Prize "for the development of the Internet."

Building the Internet was an example of phenomenal collaboration among engineers in academia, government, and the private sector. And one of the keys to this collaboration was the ability—from the earliest days of time-sharing—to use computers not just for computation but also for communication among colleagues. Because "dumb" terminals had no memory or storage, people would leave simple text messages in each other's directories on the time-sharing system, rather like leaving a note on someone's desk, which message recipients would see when they logged on. This worked fine for colleagues using the same computer but was no help for colleagues at different facilities. As soon as the first ARPANET began connecting computers over networks, users needed a way to send a message to a particular address. In the hectic months leading up to ARPANET's debut in October 1972, its developers were looking to improve communication and coordination among themselves.

Thus, in late 1971, the first "killer app" for the Internet was born—the electronic message software that we know today as e-mail. Ray Tomlinson, who worked for ARPANET contractor Bolt Beranek and Newman, wrote the first simple send-and-read software. He also devised the convention of using the @ symbol to signify sending messages from userA@computerX to userB@computerY and sent the first message to himself, from one computer to another. The two machines were side by side in the same room but connected to one another only via ARPANET.

Within a few months, others were writing software to organize and enhance e-mail features and soon e-mail made up 75 percent of all ARPANET traffic. Today it remains the most commonly used application by hundreds of millions of people around the world.

In late 1971, Ray Tomlinson used the "at" sign—now such a familiar part of e-mail addresses—and sent the first e-mail message to himself, from one computer to another over ARPANET. By 1991, the volume of traffic on the backbone and regional networks of NSFNET *(above)* was being measured in billions of bytes, ranging from zero bytes *(purple)* to 100 billion bytes *(white).*

REVOLUTION!

From Batch Processing To Time-Sharing

Using the resources of the first commercial computers was a painful process. Users keypunched program commands and data on to paper cards and then submitted the stack to a computer operator, who loaded them into the computer's card reader when nothing else was running. Your job might take only a few seconds of actual computer time, but it could be hours, or even days, before you got back your results, usually printed on green bar computer output paper. A mistake as simple as a misplaced comma meant that instead of meaningful results those pages would contain a "core dump"—an incomprehensible printout of the computer's core memory after your program failed. You then had to find the error, punch a new card, and resubmit your job.

Time-sharing took advantage of the fact that any single user made inefficient use of a computer—entering information in bursts followed by long pauses. But if many users worked at the same time, the computer could turn from one job to another during even brief pauses. Users interacted with the computer through terminals that gave them almost immediate feedback. The sense of being the only user foreshadowed the personal computer revolution to come.

Office workers at time-sharing terminals connected to a PDP-8.

Birth of the World Wide Web

E-mail was both an essential tool in the collaborative work needed to create the Internet in the first place and a new model for person-to-person communication. But for Tim Berners-Lee, a software engineer at CERN, the European Organization for Nuclear Research in Switzerland, e-mail was not sufficient.

In 1989, Berners-Lee shared the frustration of many of his colleagues at the difficulty of keeping track of experiments and information in their fast-paced world. In his observation, people responding to an article posted by one scientist might refer not just to that message or topic but also to each other's messages or topics, creating a dense web of digital information in which researchers found it increasingly difficult to locate material relevant to their own research. The best way to access and share that store of knowledge, Berners-Lee concluded, was to use the technique known as hypertext, with links that let a reader jump from the mention of a document to the document itself, allowing users to navigate CERN's huge store of information in any direction.

That March, Berners-Lee submitted a plan for "information management" to his boss at CERN, who called it "vague but interesting." Given the go-ahead to flesh out the proposal, Berners-Lee and Belgian systems engineer Robert Cailliau grafted the hypertext idea onto the Transfer Control Protocol (TCP) and Domain Name System (DNS) already in use on the Internet. The resulting Hypertext Transfer Protocol—HTTP—formed the basis for what would become the World Wide Web, which they described as a web of hypertext documents that "browsers" could view.

In December 1990, they demonstrated prototype software for a basic Web system at CERN. Each file was tagged with the prefix "http," followed by "www" (World Wide Web) and a uniform resource locator (URL) identifying the site's physical host along with the name and location of the file in the host's directory. Visitors to the first Web page—at CERN—could learn about hypertext and the Web project itself, as well as find technical details for creating their own Web pages.

Berners-Lee had set out to solve a problem for a few thousand specialists who wanted a way to access information in their own organization. But very quickly people outside the organization were accessing the Web, and in 1993 ("badgered" into it by Berners-Lee and Cailliau), CERN's directors made the Web freely available to the general public. The free Web soon outstripped a rival that charged a fee. "The whole web had always been done by people who were very interna-

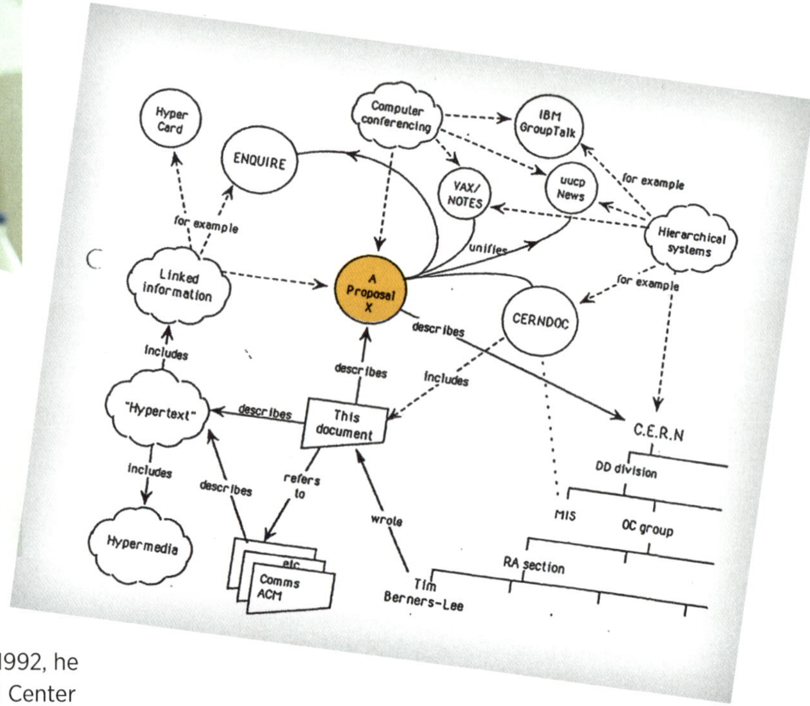

tionally-minded, very public-spirited, and very excited about the outcome," Berners-Lee would say years later. In 2007 he would be awarded the Draper Prize "for developing the World Wide Web."

Using the World Wide Web requires software called a browser to retrieve, present, or navigate information resources on the Web. Berners-Lee's browser was called WorldWideWeb because at the time it was the only way to see the Web. "Much later it was renamed Nexus," he would recall, "in order to save confusion between the program and the abstract information space (which is now spelled World Wide Web with spaces)."

The original Web and browser dealt only in text. One of the earliest browsers to introduce static images (video was still years away) was developed by a young computer whiz named Marc Andreesen, an engineering student at the University of Illinois at Urbana-Champaign. During his senior year in 1992, he took a part-time job at NSF's National Center for Supercomputing Applications, where he gained access to the World Wide Web. Before Andreessen graduated, he and others at the center, including fellow student Eric Bina, devised a graphically enhanced Web browser called Mosaic, which was released free over the Internet in 1993. Mosaic proved hugely popular and led to an explosion in Web use.

Jim Clark, a successful computer programmer and entrepreneur, then teamed with Andreessen to launch Netscape, a company that adapted Mosaic for commercial purposes. Netscape Navigator was released in 1994 and was the dominant Web browser for that decade. Among Netscape's innovations were so-called cookies, which track visits to websites, allowing advertisers to identify user interests, and a technique for encrypting credit card numbers so that purchases could be made safely over the Internet. Other innovative "dot-com" companies founded in the 1990s included the Internet retailer Amazon and the search engine Google, which intended to make all information publicly available over the Internet. Both companies would expand far beyond the United States, relying heavily on engineering advances in data storage to pack ever-increasing amounts of data into ever-shrinking hard drives.

The Internet, the World Wide Web, and search engines promoted globalization and the rewards and risks that came with it. Today the term "web" describes more than just the cyber universe of information resources. It is an increasingly apt description of how the world is knitted together.

Tim Berners-Lee, shown at CERN with the NeXT computer he used to invent the World Wide Web, wrote his revolutionary proposal for the Web in March 1989. The cover of the proposal *(above)* sketches how hypertext links would allow users to follow their interests from source to source.

WORKING IN SPACE

Still Above and Beyond

The period from the late 1960s to the early 1990s was a time of extraordinary engineering accomplishments in the space program. Following the successful Apollo moon landing, aerospace engineers transformed the space program from a series of one-time launches to a mature program with frequent flights into space. That work gave mankind a permanent orbiting laboratory at the International Space Station (ISS). As of 2014, the ISS has been continually occupied for 14 years.

Five U.S. Space Shuttles flew 135 missions, collectively spending more than three and a half years in orbit. Dozens of these were in support of the ISS, but others performed important science. One shuttle experiment demonstrated that dangerous bacteria get even more dangerous in low gravity. By investigating the process involved, biomedical researchers have been able to develop a *Salmonella* vaccine. Research with other disease-causing microbes may lead to similar breakthroughs in the future. In the course of the 135 missions, two shuttles were lost, indicating what a daunting systems engineering challenge going into space was and still is.

One of the most exciting and productive eras in space-based science began with the deployment of the Hubble Space Telescope in 1990. Earthbound engineers worked with astronomers to build an instrument that would spend decades in orbit to give us pictures of the furthest corners of the universe. Hubble has given us a close look at the planets within our Solar System, and found planetary bodies orbiting distant stars. Its observations have also shown that the expansion of the universe is accelerating, which astronomers now believe is evidence of "dark energy."

The ISS, the Hubble Space Telescope, and many smaller probes and spacecraft embody the engineering that has made it easier to reach outer space while producing technologies that benefit the Earth below.

The space shuttle *Endeavour*, docked at the ISS, as viewed from a departing Russian Soyuz spacecraft.

Soaring prices and long lines for gasoline in the early 1970s led to federal requirements doubling average fuel mileage to 27.5 miles per gallon within 10 years.

Clean and Efficient Energy

By 1989, engineers had made notable progress in designing motor vehicles that were both less polluting and more fuel-efficient. Those advances were prompted by growing awareness of the health risks of air pollution and the need for Americans to conserve oil and gasoline, which was largely imported and increasingly expensive. Actions by the Organization of Petroleum Exporting Countries (OPEC) in the early 1970s to restrict supply had caused U.S. gasoline prices to soar, prompting Congress in 1975 to require that the average fuel efficiency for cars be doubled to 27.5 miles per gallon within 10 years. At the same time, clean air laws went into effect on new cars that strictly limited harmful exhaust emissions, including carbon monoxide and other compounds that can cause serious damage to human health and the environment. States required periodic inspections of cars to ensure that they met tailpipe emissions standards.

Automotive engineers responded to those challenges by developing new techniques and improving devices that processed vehicle exhaust. Building on the 1950s pioneering work of French engineer Eugene Houdry, who had used catalysts to turn unburned hydrocarbons from car exhaust into carbon dioxide and water, American engineers created the first practical commercial, two-way catalytic convertors in the 1970s to reduce hydrocarbon and carbon monoxide emissions. In 1981 an engineering team led by Carl Keith and John Mooney at Engelhard Corporation designed the three-way catalytic converter, still used today to reduce auto emissions of carbon monoxide, hydrocarbons, and nitrogen oxide (a gas that contributes to smog and acid rain). The three-way catalytic converter significantly improved public health. As one EPA official said when Keith died in 2008, "Billions of people around the world breathe cleaner air because of this invention."

Carl Keith *(far left)* and John Mooney received the 2002 National Medal of Technology and Innovation from President George W. Bush for the invention of the three-way catalytic converter.

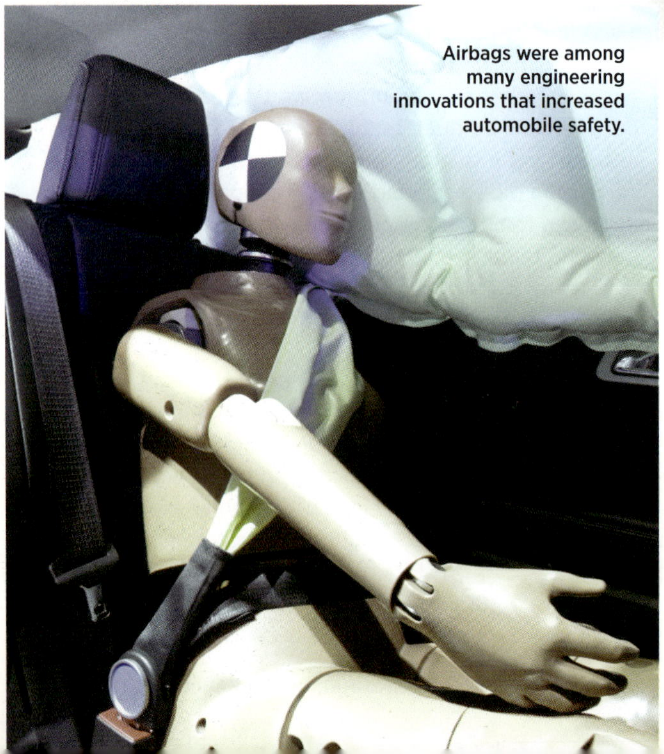

Airbags were among many engineering innovations that increased automobile safety.

Increasing mileage to meet government standards required further engineering advances, including the use of sturdy but lightweight construction materials such as aluminum, duralumin (a strong aluminum alloy), engineered plastics, and fiberglass. Reductions in weight tend to make cars less crash-resistant, but engineers compensated with design techniques that improved crash tolerances. Along with other safety measures, such as road safety engineering and speed regulations, the incorporation of devices such as seatbelts, airbags, antilock brakes, and running lights helped make driving less dangerous, even as autos became lighter on average and more fuel-efficient. Between 1970 and 1990, U.S. traffic fatalities decreased 57 percent, from an annual rate of 4.85 deaths per 100 million vehicle miles traveled (VMT) in 1970 to 2.08 deaths per 100 million VMT in 1990.

To meet clean air standards, engineers had to reduce pollution not just from cars and trucks but also from other major sources. Homes, offices, apartment buildings, and hospitals produce their share of emissions, mainly from heating. More significant sources include industries and power plants, many fueled by coal, which often contains significant amounts of sulfur. Historically, the combustion of coal in industrialized countries produced both major economic benefits and serious health and environmental risks, creating smog so thick in some cases that it proved deadly. Environmental concerns and laws spurred engineering solutions to those problems. In 1985 the Department of Energy (DOE) launched the Clean Coal Technology Program, which sponsored research that made coal burners more efficient and reduced emissions of pollutants. Existing technologies such as coal scrubbers were improved and installed at many power plants, where they captured significant amounts of sulfur dioxide, the major contributor to acid rain, before the exhaust was released into the atmosphere. Emissions of nitrogen oxides were also reduced using improved coal burners and scrubbers. Since 1990, those advances—combined with cleaner exhaust from cars, trucks, and other sources—have helped cut annual emissions of sulfur dioxide in the United States by 75 percent and annual emissions of nitrogen oxides by 50 percent.

By the late 1980s, another environmental challenge was emerging for engineers and for the world at large. The Clean Air Act more than

By 1992, 110 nuclear power plants were contributing nearly 22 percent of electricity produced in the United States—a figure that has changed little since.

two decades earlier had addressed the generation of pollutants such as sulfur dioxide, nitrogen oxides, and particulates from the burning of fossil fuels. Now atmospheric accumulations of greenhouse gases carbon dioxide and methane were linked to a disturbing long-term increase in global temperatures. (Greenhouse gases are so named because the historical record shows that their excess presence in the atmosphere blocks the escape of heat from the planet.)

On February 22, 1989, Rep. Claudine Schneider of Rhode Island introduced the Global Warming Prevention Act, which called for the United States to reduce carbon dioxide emissions 20 percent by 2005. According to the U.S. Office of Technology Assessment, carbon dioxide emissions (as carbon) from energy use—including gasoline for automobiles, natural gas for home heating, and various fuels for generating electricity—totaled about 1.4 billion metric tons in the United States in 1989, roughly 20 percent of the world total.

Regulations to reduce emissions would have required disruptive changes in energy systems across the country, however, and many political leaders resisted. Even with bipartisan support from 144 congressional cosponsors, the Global Warming Prevention Act did not get out of committee and to the House floor.

Although reducing carbon emissions was a nonstarter, efforts to find cleaner sources of energy had been under way at DOE since the oil crisis in the 1970s, when several national laboratories began to research alternative, largely naturally renewable sources such as solar, wind, hydropower, and geothermal power. For example, the National Renewable Energy Laboratory (NREL, established in 1977 as the Solar Energy Research Institute) in Golden, Colorado, has long conducted research and promoted development of solar energy and other renewable sources. Solar photovoltaic power has been enhanced through development of more efficient solar cells and of engineering systems that use lenses to intensify the sunlight charging the cells. Two solar thermal demonstration projects were launched in California's Mojave Desert. Solar One (1982) and Solar Two (1996) used mirrors to construct a solar amplifier and focus sunlight on receivers in thermal towers. The receivers stored heat in a liquid medium and then used it to produce steam to power turbines and generate electricity—methods now used to provide electricity to power grids. Research by the NREL also helped engineers design modern wind turbines that generate 15 times more electricity than did the average turbine in 1990.

Nuclear power plants, although not a source of renewable energy, produce no carbon dioxide or other air pollution. In 1979, 72 licensed reactors produced 12 percent of the nation's electrical output. Responding to public

Technology first developed in the mid-1990s at projects such as Solar Two *(above)* in the Mojave Desert demonstrated that solar heat stored in thermal towers could be used to produce steam to power turbines to generate electricity.

concerns about the safety of nuclear power following a partial meltdown at Three Mile Island in Pennsylvania in 1979, the U.S. Nuclear Regulatory Commission and the industry's Nuclear Energy Institute raised design standards for reactors and improved training and emergency-response measures. Engineers met the new design standards, and by 1992, 110 nuclear power plants were contributing nearly 22 percent of electricity produced in the United States—a figure that has changed little since.

Building for Safety

Around 5 p.m. on October 17, 1989, an earthquake measuring 6.9 on the Richter scale shook the San Francisco Bay area, convulsing a region inhabited by more than 6 million people. The quake caused more than 60 deaths, injured nearly 4,000 people, and caused $6 billion in damage to buildings, roads, and bridges. Still, the earthquake indicated how far engineering had progressed in designing for disasters. Except for old masonry structures or those situated atop loose, sandy soil that liquefied, the vast majority of buildings remained unaffected. San Francisco's Candlestick Park, where fans were awaiting the first pitch of the third game of the World Series between the hometown Giants and the Oakland Athletics, was shaken sufficiently to result in an emergency evacuation but remained intact. And the city's tallest building, the Transamerica Pyramid—constructed in the early 1970s with massive concrete-and-steel trusses at its base to withstand seismic shocks—swayed during the quake but suffered no damage.

Historic Oakland City Hall was retrofitted after the 1989 earthquake using base isolation techniques that placed 112 rubber and steel bearing pads between the building and its foundation.

During the 1970s and '80s, an effective way to buffer buildings of average height against earthquakes was developed collaboratively by structural engineers such as William Robinson of New Zealand and James Kelly of the University of California, Berkeley. The technique, known as base isolation, involves placing bearings made of rubber or other shock-absorbing materials between the ground and the base of a structure. Oakland City Hall *(right)* was among the older buildings retrofitted in that manner following the 1989 earthquake. And, in January 1994, a new building incorporating the same technology—the USC University Hospital in Los Angeles—would fare well during a 6.7 magnitude earthquake that damaged other hospitals in the area and caused patients to be evacuated.

To avoid loss of life and mitigate the steep cost of retrofitting buildings, structural engineers developed even more precise ways of predicting and countering the potentially deadly impact of seismic shocks and other stresses on buildings before they are built. In many buildings in seismically active zones, engineers installed instruments to record how the buildings respond to tremors large or small. With each earthquake, fresh data went into computer models that use the finite element method (FEM) to predict how designs still on the drawing board would fare in that quake. Engineers began testing scale models of small buildings on "shake tables" that simulate an earthquake using earlier recordings of earthquake ground motions. They also developed probes to determine whether soil would liquefy during severe tremors and where construction should be avoided or existing structures should be reinforced. Such safety measures could save countless lives when high-risk zones like the San Francisco Bay area suffer major earthquakes in the future.

James Watson was named first director of the Human Genome Project to determine the order of the nucleotide base pairs that connect the twin strands of DNA that make up our chromosomes.

Knowing Ourselves: Biomedical Engineering

In 1953, molecular biologist James Watson and British biophysicist Francis Crick determined the double helix structure of the DNA (deoxyribonucleic acid) that makes up our chromosomes, the structures in the nuclei of our cells containing the thousands of genes in the human genome. That breakthrough led to feats of genetic engineering with far-reaching benefits. For example, in the 1980s, recombinant DNA technology—which splices together strands of DNA from different species to produce genetic sequences not found in nature—yielded new strains of disease-resistant crops. It also led to the creation and manufacture of a form of human insulin that is less likely to cause allergic reactions when administered medically than earlier forms of insulin extracted and purified from the pancreas of pigs or cows.

As gratifying as these first uses of genetic engineering were, medical researchers believed that deciphering the human genome would have even more profound consequences for human health care. Decoding the genome could lead to treatments for many of the more than 4,000 genetic diseases that afflict humanity, as well as for disorders in which genetic predisposition is important. For example, if DNA sequencing reveals a genetic predisposition to certain forms of cancer, frequent diagnostic testing could lead to early detection of the disease and an opportunity for effective treatment.

Thus, in November 1989, the NIH established the National Center for Human Genome Research, choosing James Watson to be its first director. Watson's job was to launch the massive Human Genome Project to determine the precise order, or sequence, of the nucleotide pairs that link the twin strands of chromosomal DNA like the rungs of a ladder. The base pairs, as they're called, come in two combinations—*A* and *T* (adenine and thymine) or *C* and *G* (cytosine and guanine)—and the genome contains about three billion of them.

The U.S. government took the lead for the Human Genome Project by investing nearly $3 billion, but it was defined from the start as a global research effort. The process of sequencing human DNA originally involved much laborious manual effort, and skeptics warned that the project might take several decades to complete and cost tens of billions of dollars. But very soon, innovative engineering—such as using robotic arms to perform meticulous and repetitive lab procedures—helped researchers advance toward the goal of fully automated sequencing techniques, allowing them to transcribe and decipher the base pair code much more quickly and efficiently than predicted.

Because sequencing could be done only on short bits of DNA, the results were like tiny pieces of an immense puzzle that researchers assembled and interpreted using supercomputers. Among those who refined such techniques were Francis Collins, who would succeed James Watson as director of the Human Genome Project in 1993, and J. Craig Venter, who founded Celera Genomics, a company that used a controversial "shotgun" method that expedited sequencing. The public-private competition, which involved interdisciplinary teams and automated procedures at a few major centers, would help complete the process by 2003, two years ahead of schedule.

In a fitting nod to the digital age, the full genome sequence was published on the

J. Craig Venter (*below*) founded Celera Genomics to find faster ways of sequencing the human genome. Instruments like the one at left are used to break DNA into small pieces for analysis.

Internet rather than in book form. Had the sequence been committed to paper, it would have filled hundreds of thousands of pages with text consisting of four letters (*A, C, G,* and *T*), repeated in seemingly endless combinations—a cryptic code containing the secrets of life.

One of the project's significant findings was that the human genome contains only 20,000 to 25,000 genes—dramatically fewer than the 100,000 genes estimated a decade earlier, and fewer than the 30,000 to 35,000 genes estimated from the rough draft of the genome finished in 2000. According to Project Director Collins, "The availability of the highly accurate human genome sequence in free public databases enables researchers around the world to conduct even more precise studies of our genetic instruction book and how it influences health and disease." The creative engineering that speeded the laborious task of mapping the human genome also helped make DNA analysis for individuals more efficient and affordable, allowing people to research their ancestry, for example, for a few hundred dollars or less.

The 1980s also saw breakthroughs in orthopedic biomechanics and biocompatible materials to create artificial hip and knee replacements, as well as in methods of using electromagnetic signals from muscle contractions to control prosthetic hands, arms, and legs. New electronic devices—implanted defibrillators—were developed to automatically restore normal heart rhythms to individuals experiencing otherwise fatal fibrillation. Breakthroughs in tissue engineering led to artificial skin made from collagen, silicone, and other substances being used surgically to treat those who had suffered severe burns, but sometimes the patient's immune system rejected such grafts. That problem was addressed by Eugene Bell of MIT, who founded a company called Organogenesis that produced Graftskin, which included cultured human cells and proved successful in clinical trials. In 1998 Graftskin would become the first living engineered tissue approved for use by the Food and Drug Administration (FDA). Other tissues engineered for medical purposes included bone, cartilage, and even arteries.

Before the end of the 20th century, medical advances that earlier generations would have dismissed as fantasy—the creation of new organs and organisms in laboratories—had become possible. Engineering advances not only in biomedicine but also in public health and safety, energy efficiency, and digital computing and communications had transformed the world in ways that few of those alive at the start of the century could have imagined.

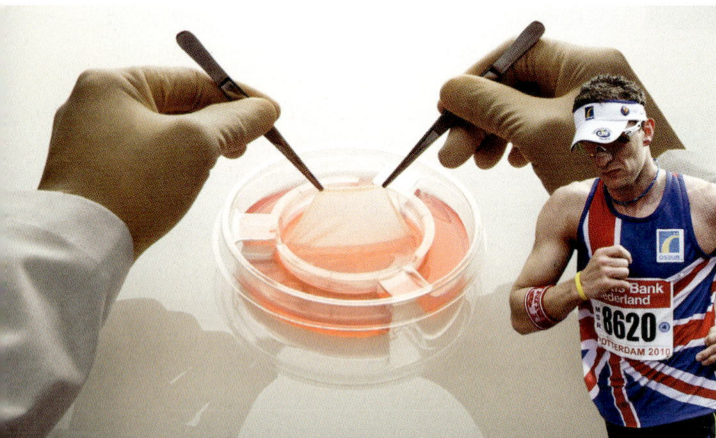

Breakthroughs in the 1980s in biocompatible materials and orthopedic biomechanics have led to artificial skin, as well as to artificial hip and knee replacements. Meanwhile, with new methods of using electrical signals from muscle contractions people can control prosthetic arms, hands, and legs.

2014 A Healthier, Cleaner, More Connected World

Surrounded by the fruits of innovation, we easily forget that much of what we now take for granted in 2014 was almost unimaginable 25 years ago. Back in the 1980s, pagers, e-mail, and floppy disks were cutting-edge technologies. Today we have smartphones, flash drives, and "the cloud." Back then the idea for the World Wide Web was just beginning to germinate in the mind of Tim Berners-Lee. Today roughly 40 percent of the world's population uses it, each of us for our own purposes. We go to the Web to learn and to share, to buy and sell, to meet new people and locate old friends, to check the weather, pay bills, and renew the car registration. And the more we use it, the more it evolves to meet our needs.

Technological advances made over just a few decades are boosting economies, feeding the hungry, and healing the sick. Iowa farmers achieve record yields with gene-spliced crops and other agricultural technologies. New vaccines hold out the promise for tackling scourges like malaria and some cancers, while doctors save lives by replacing diseased heart valves—in some cases without open-heart surgery. And who would have thought that in 2014 simple robots would vacuum our houses, highly complex ones would assist surgeons in performing lifesaving surgery, and cars would automatically slam on the brakes when a child darts out in front of them?

All these advances have come through engineering carried out in companies, universities, and national laboratories. Those efforts have created new materials like nanotubes and high-strength alloys, manufacturing technologies like 3-D printing, software and algorithms for harnessing the power of supercomputers and mining vast stores of data, and countless other innovations.

Yet these examples barely scratch the surface of the remarkable changes wrought over the last quarter century. Our lives, our workplaces, our societies have been transformed by an extraordinary flowering of engineering innovations. Life offers more possibilities, more richness, than ever before.

Wind turbines offer a clean, renewable source of energy.

Introduced in 1971, the Intel 4004 microprocessor *(left)* contained 2,300 transistors. The exponential increase in transistor counts *(see chart opposite page)* on tinier and tinier chips has led to such modern devices as digital cameras that can capture a breaking wave in mid-air.

Tiny Powerhouses, Global Reach

By the early 1980s, the semiconductor revolution was well under way. In 1982, engineers were packing 134,000 transistors on a single microprocessor chip, making the personal computer possible. In 1985, that number jumped to 275,000. But the chip designers began to run into a physical limit on the number of transistors on a chip. As the transistors got smaller and smaller, the width of each individual component in a chip design began to approach the wavelength of the visible light being used to transfer the design onto a silicon crystal wafer in a process called photolithography. As a result, the features—the transistors and connecting wires—weren't printed precisely enough to operate reliably. The features would get fuzzy, instead of being sharply delineated, allowing short circuits and causing the chips to fail.

The solution, IBM electrical engineer Kanti Jain realized, was a lithography tool that used shorter wavelength deep ultraviolet light instead of visible light. Jain and his team tapped into a device invented by Russian engineers in 1970s—the excimer laser, which creates ultraviolet light with electrical stimulation and high pressure on gas combinations such as krypton and fluorine. But the prevailing wisdom held that lasers would never work for lithography. Jain had to develop the complex optics to evenly illuminate the silicon wafer with the laser and to engineer a wafer coating of photosensitive material, or photoresist, that responded to ultraviolet light. In 1982 he succeeded—and within a decade, the big semiconductor manufacturing equipment companies offered commercial ultraviolet lithography tools, or steppers, capable of executing chip designs at the high resolution necessary.

It's hard to overstate the importance of myriad essential engineering advances like this one in semiconductor manufacturing. The innovations kept Moore's Law—the idea that the number of transistors on a chip doubles every two years—from hitting a wall. Now, commercially available microprocessors contain more than 7 billion transistors, packing more than 8.75 million on every square millimeter. Each individual feature is only 22 billionths of a meter wide—4,000 of them side by side span the width of a human hair. The consequences have been profound.

Without the vast increase in transistors on a chip and the resulting huge leap in computing power—combined with the complex software needed to unlock that power and the falling costs that have made products accessible—we'd have no super-realistic video games or our now-essential smartphones. Supercomputers wouldn't be modeling weather patterns and dangerous storms days in advance, accurately predicting blizzards in Colorado and floods in Bangladesh—and saving countless lives. Companies wouldn't be doing most of the design work for fuel-efficient airplanes through simulations alone.

Now, thanks to sophisticated silicon chips, autonomous submersibles can chart ocean currents to monitor the health of the oceans and answer questions about climate change. Flying drones track orangutans in Indonesia and ivory poachers in Africa.

Computers on the electrical grid balance supply and demand, ensuring that the lights stay on in our homes and factories keep humming. Meanwhile "smart" meters enable the solar panels springing up on hundreds of thousands of roofs to feed clean power back into the grid.

Or consider another science and engineering breakthrough. Physicists had known since 1856 that the resistance to electrical current flowing through many metals changes slightly in a magnetic field, a phenomenon called magnetoresistance. In 1988, French physicist Albert Fert thought he could amplify the effect by designing materials made up of very thin layers of metals. He tried sandwiching chromium with iron—and achieved a magnetoresistance 10 times that of standard metals. About the same time, German physicist Peter Grünberg independently managed a similar feat.

MICROPROCESSOR TRANSISTOR COUNTS FROM 1971-2011 AND MOORE'S LAW

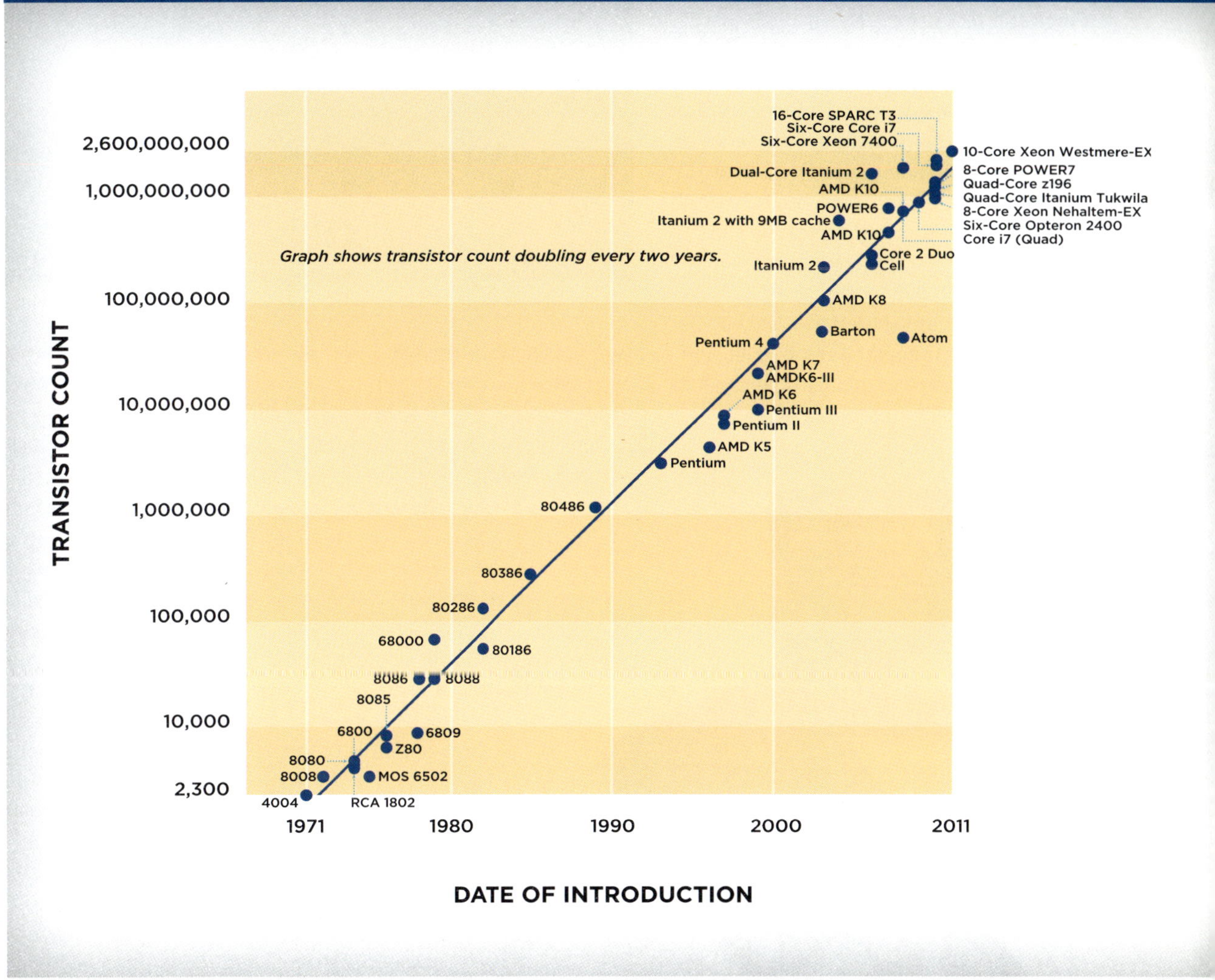

Computer disk drives already depended on magnetoresistance to read data stored magnetically on spinning discs. But Fert's and Grünberg's "giant" magnetoresistance promised dramatic gains in storage density—if a series of complex engineering and materials science problems could be solved. They could. By 1994, IBM engineers had produced prototype hard disks that stored 17 times more information per square inch than previous devices.

Fert and Grünberg shared the 2007 Nobel Prize in physics. By then, their discoveries and subsequent advances had enabled the capacity of data storage devices to double every year—even faster than Moore's Law. Now, what once seemed like science fiction is part of daily life: Movies on demand. Entire libraries of books or music in the palm of your hand. Maps and photographs of virtually every street in the United States—and many countries around world—accessible at a keystroke. With big data centers and complex software, companies now manage and control vast supply and distribution chains, track customers' purchases and preferences, and offer unprecedented levels of personalized services.

Yet smaller transistors and expanding data storage are just a tiny fraction of the engineering wizardry that has transformed our lives. Light-emitting diodes and liquid crystal displays have made flat-panel video screens ubiquitous, from living rooms and stadiums to myriad handheld devices and heads-up displays. Glass fibers now carry terabytes of information around the world in flashes of light. GPS guides airplanes, farm tractors, ships at sea, and ordinary travelers in their cars or through their cell phones. Software and algorithms make sense of huge databases and connect people through social media. Cellular phone networks offer instant connections, even from distant mountaintops. And lithium-ion batteries provide hours of energy to run our cell phones, laptops, tablets, cameras, cordless power tools, and many other compact, lightweight mobile devices. The creators of fiber optics, lithium-ion batteries, GPS, cell phone networks, charge-coupled devices (the sensors in digital cameras), and liquid crystal displays—innovations that are now integral to life in the 21st century—have all been recognized by the nation's top engineering award, the National Academy

With computer chip "brains" an autonomous underwater vehicle can record the effect of ocean currents on fish larvae in the waters off Belize *(above left)*. Equipped with GPS technology, drivers can find their way in unfamiliar cities *(above)*.

Peter Grünberg *(left)* and Albert Fert independently developed "giant" magnetoresistance.

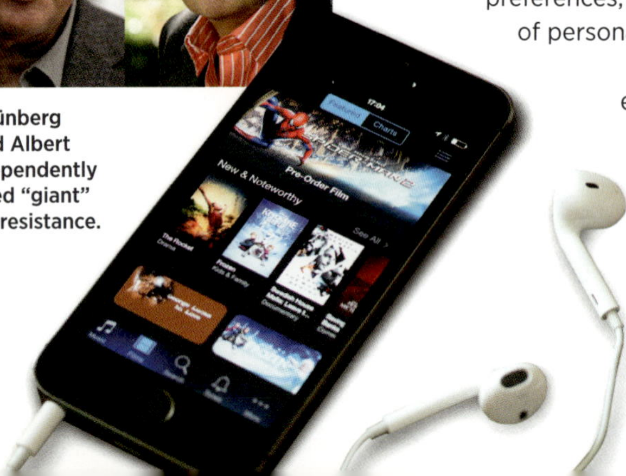

of Engineering's Charles Stark Draper Prize for Engineering.

Together, these advances have created a connected world rich in information that is expanding at a breathtaking clip. In 2014, we use smartphones to book flights, pay bills, hold meetings, and settle arguments at the dinner table. In Nigeria, farmers who never heard a phone ring as children now use cell phones to line up customers and get vouchers for seed and fertilizer. Governments connect directly with citizens, giving the public access to valuable data on contracts and spending, and taking action on complaints. With a computer and an Internet connection, a soldier fighting a distant war can sing his three-year-old daughter to sleep as she watches on the home computer screen. High-speed connections have become as important for economic growth and commerce as railroads and highways once were, as cities like Chattanooga, Tennessee, attract new companies with broadband networks. People living in Beijing, Berlin, and Boston can work together almost as if they were in the same room.

Meanwhile, social media are creating communities and connecting friends, while also becoming a potent political tool. In the week before Egyptian president Hosni Mubarak resigned in 2011, for instance, the number of tweets about political change in Egypt climbed to 230,000 a day. Protest and political videos went viral, with millions of views. In the Arab Spring, concluded a University of Washington study, "social media carried a cascade of messages about freedom and democracy across North Africa and the Middle East, and helped raise expectations for the success of political uprising."

Of course, this connectivity isn't necessarily all sunshine and roses. For instance, many would say that the Arab Spring has failed to deliver on its initial promise of spreading freedom and democracy. The new world also comes with thorny new problems. Operating around the clock, global engineering enterprises, including manufacturers of everything from clothing, cell phones, and computers to appliances, automobiles, and aircraft, can assemble talent globally—and also outsource IT tasks and other jobs from high-wage countries like the United States to lower-wage developing countries.

By breaking through the security walls of company databases, thieves have been able to steal credit card information and other valuable data. Terrorists, like every other kind of organization, have become adept at using the Internet to communicate and plan. Many people are overwhelmed by the flood of seemingly urgent e-mails and by information in general. Meanwhile, fierce debates are raging about governments spying on the communications of their own citizens, about companies collecting vast amounts of information on people's online habits and behavior, even about whether to use cameras to monitor traffic and fine drivers remotely for running red lights or speeding.

The good news, though, is that many of these problems will likely be solved by more innovation. Where necessary, government policy and regulation can address many issues as well, in a feedback loop that relies on continued engineering creativity.

Whether chatting with a family member on the other side of the world or holding business meetings with far-flung colleagues, people communicate via computers wherever they can connect to the Internet.

Bald eagles came back from the brink of extinction thanks to the elimination of chemicals like DDT from the environment.

A Cleaner World

A whole generation in the United States has grown up without memories of rivers catching on fire, smoke darkening Pittsburgh, smog hanging over Los Angeles, or of acid rain rendering one-quarter of the lakes in the Adirondacks too acidic for fish to live. They don't remember that on cold winter mornings across much of America, the air was heavy with the stomach-churning odor of unburned hydrocarbons as people cranked their car engines.

Below left: **Mark Musculus and colleagues at Sandia National Laboratories use optical diagnostic techniques to identify pollutants in motor vehicle fuel.**

Although we still face daunting environmental challenges—oil and chemical spills, nutrient pollution in rivers and lakes, ocean acidification, habitat loss, and species extinction—consider how much progress has been made. On sunny days, cities now sparkle under blue skies most of the time. For the most part, rivers and lakes are clean enough for swimming. The ozone hole is closing, and the burden of lead in our bodies is dropping. Eagles and many other species have rebounded from the chemicals that almost caused their extinction. Americans even use less energy per person—about 10 percent less—than they did in 2007. Many of these improvements—spurred by governmental policy and public demand—required new technologies created through engineering innovation.

Researchers at Sandia National Laboratories and other labs, for example, worked with auto and truck companies to create ways to burn motor vehicle fuel more cleanly and efficiently. That step forward has been especially important for diesel engines, whose exhaust is harder to clean with catalytic converters than that of gasoline engines. Add in a host of other innovations from the auto industry, such as variable valve timing, direct ignition, and up to a hundred microcomputers in a single car, and today we have vehicles that are more than 95 percent cleaner than those in the 1960s. The plumes of black smoke once billowing from trucks plying the nation's highways have mostly vanished. Cars and SUVs are much safer and more powerful than in decades past, and packed with features like power windows; antilock brakes; air bags; and a number of safety, comfort, and handling control and convenience features. Yet average fuel economy has climbed to 24.8 miles per gallon in 2013 for cars and light trucks, up from 20.8 mpg in 2008 and far above the 1975 level of 12.9 mpg. Many models achieve more than 40 mpg.

The result? Look at Los Angeles as just one example, where the number of health advisories from unhealthy ozone-laden air dropped from 144 in 1988 to 0 in 2012.

Moreover, the pace of innovation in automobiles and other vehicles continues to accelerate. Buyers can now choose from more than two dozen hybrid models, the most efficient of which are rated at 50 mpg combined city and highway driving running on gasoline. Consumers can also select from more than a dozen all-electric models, with companies and states racing to build charging stations on major highway routes so that owners won't suffer from "range anxiety"—the fear of running out of juice. On January 30, 2014, two electric cars headed out across the entire United States from Los Angeles to New York City, enduring blizzards, freezing temperatures, a blinding sandstorm, and driving rain. They made the journey in about 76 hours—including about 15 hours of charging time.

Moreover, the pace of innovation in automobiles and other vehicles continues to accelerate.

Energy efficiency has been and remains the low-hanging fruit for reducing reliance on fossil fuels, and engineers have done much in recent decades to conserve fuel and reduce pollution by designing energy-efficient buildings. "Green" roofs both help reduce the urban heat-island effect and cut pollution from storm-water runoff that contaminates rivers and streams. Many energy-saving technologies, including coated glass windows that conserve interior heat in cold weather and deflect exterior heat in warm weather, were engineered at the Lawrence Berkeley Laboratory in California under the direction of Arthur Rosenfeld, a particle physicist by training who went on to become a member of the California Energy Commission. The efforts of the commission, which included pioneering efficiency regulations, have helped reduce the amount of energy used per person in California to a level about 40 percent below the nationwide average.

Similar progress has been made on many other fronts. Refrigerators, TVs, and computers are far more energy-efficient than they were a quarter century ago because of engineering advances and new efficiency standards. Tollbooths are being replaced by automated toll collection systems that speed travel as well as reduce pollution.

Companies have harnessed a technology called cavity ring-down spectroscopy to engineer mobile methane detectors that can spot leaks of natural gas, which consists primarily of methane, a greenhouse gas.

Farmers use yield sensors, autopilot-guided tractors, variable computer-controlled sprayers, and other recent advances to apply just the needed amount of fertilizer and water to each small patch of their field. The technology saves money and water, boosts yields, and reduces the nutrient runoff that flows into rivers and lakes, another environmental concern. Along with polluted runoff from parking lots and pavements, nutrient runoff has triggered blooms of toxic algae and created vast "dead zones" that kill marine life over thousands of square miles in the Gulf of Mexico.

Keeping the air and water clean, protecting people from chemicals and toxins, and reducing greenhouse gas emissions are never-ending struggles. Engineers aren't the only troops in these continuing battles. But the solutions they create have been—and will continue to be—the essential part of any victories that we achieve. ●

Two all-electric Tesla Model S vehicles drove from Los Angeles to New York City in the middle of winter. During one overnight leg the team endured more than 12 inches of snow, icy roads, and high winds. The only breakdown occurred with one of the gasoline-powered support vehicles.

HYBRIDS

An Idea Whose Time Arrived at Last

Strictly speaking, hybrids—vehicles that can use more than one form of energy—aren't new. Most long-haul railroad locomotives are hybrids, with a diesel generator that provides power to massive electric motors. And the first hybrid automobile actually dates back to 1900, when Ferdinand Porsche, working for carriage builder Jacob Lohner & Co. in Vienna, Austria, used two gasoline generators to drive electric motors built into the vehicle's wheel hubs *(above)*. Despite constant refinements to the design, very few Lohner-Porsche hybrids were made or sold. The idea was simply too far ahead of its time.

Now fast-forward 100 years to August 2000, when the 2001 Toyota Prius hybrid began arriving in dealer showrooms in the United States. With EPA mileage ratings of 52 mpg, the Prius was named Best Engineered Car of 2001 by the Society of Automotive Engineers. In 2002, Prius sales topped 100,000 worldwide. The first plug-in hybrid, the Chevrolet Volt, and first all-electric vehicle, the Nissan Leaf, arrived in December 2010.

Hybrids come in two main types. In a "series" hybrid (like the Lohner-Porsche) there is only one path to power the wheels—namely, an electric motor that gets its electricity from either high-capacity batteries or an onboard generator typically fueled by gasoline. The generator only runs when the batteries are low on power. The gas engine/generator recharges the batteries, which are also recharged through regenerative braking—capturing energy normally lost during braking and using the electric motor as a generator to store it in the battery.

A "parallel" hybrid has two complete power trains—usually a gas-fueled internal combustion engine and a battery-powered electric motor—that can work individually or together to turn the wheels and move the car. A parallel hybrid switches between the systems to get the greatest efficiency. As with a series hybrid, the battery is charged by the gas engine/generator and by regenerative braking.

Plug-in hybrids, which may be either serial or parallel, have the added ability to charge their batteries by an outside power source. They also have larger battery packs than regular hybrids, making it possible to drive using only electric power.

In 2012, with gasoline prices averaging $3.60 per gallon (and pushing $4 in some places), Americans bought more than 50,000 plug-in electric vehicles. In the first half of 2013, as battery costs were dropping—and with so many more hybrids and all-electric cars to choose from—Americans bought double the number of plug-in electric vehicles they purchased in the same period in 2012.

Ferdinand Porsche would no doubt be pleased to see his idea finally catching on.

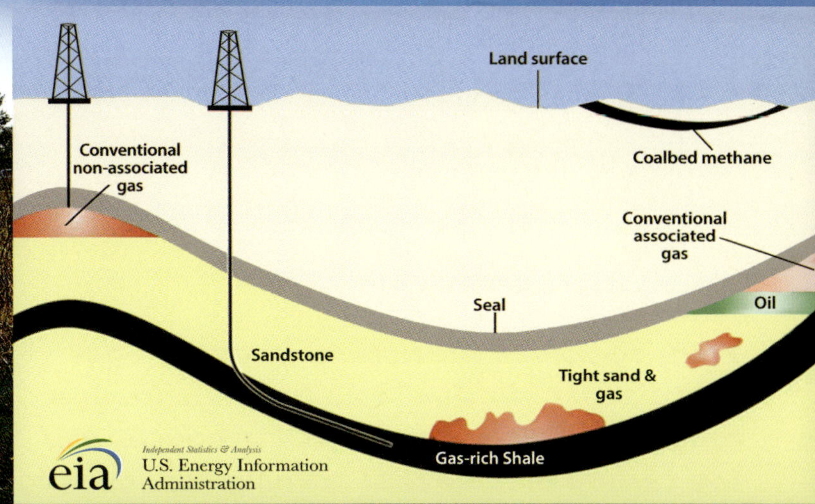

An Unexpected Change in the Energy Landscape: The Shale Gas Boom

In the 1970s, some warned that the world was on an unsustainable path and that before too long we'd run out of oil and food. Lights would dim, factories would slow, people would starve, civilization would crumble. The oil crisis of the mid-1970s, with long lines of cars waiting for gasoline, seemed to be a harbinger of that grim future. But advances in hybrid crops have sustained food supplies and, for the United States at least, new supplies of oil and gas appeared seemingly out of nowhere due to engineering innovation.

As shown in the schematic at *far right above*, horizontal drilling provides greater access to natural gas trapped deep in a shale formation. First, a vertical well is drilled to the desired depth. Then the drill bit is turned to bore a well horizontally through the reservoir.

In 1997, Mitchell Energy was in trouble. Production from the company's gas wells in Texas was falling. Reserves were declining. So founder George Mitchell took a gamble. Bucking the conventional wisdom—and the advice of top executives in his own company—Mitchell decided to step up drilling in the Barnett Shale.

Geologists had long known that shale formations deep underground contain large amounts of gas and oil. But freeing the gas trapped in the rock is difficult. Pumping down thick, viscous liquid under high pressure can fracture the rock (a process called hydraulic fracturing, commonly known as "fracking") and liberate the fuel. Although hydraulic fracturing itself dates back to the 1940s, it wasn't working in the shale formations, where production at most wells that used the technique quickly petered out.

The job of successfully extracting gas from shale fell to Mitchell engineer Nicholas Steinsberger. He experimented with different liquids and gels, with little success. Then a contractor accidently pumped down fluid that was more watery than usual—and more gas than expected came up. Could mostly water be the answer? Steinsberger decided to find out.

"Most everyone thought Steinsberger was out of his mind," wrote Gregory Zuckerman in his 2013 book, *The Frackers: The Outrageous Inside Story of the New Billionaire Wildcatters.* He wasn't. With his watery fracturing fluid, his wells kept producing and producing and producing.

Steinsberger's innovation was a key piece of the puzzle of how to tap into the nation's huge deposits of shale gas and oil, but it built on numerous other engineering advances. The most important was figuring out how to drill deep and then turn the bit sideways to drill horizontally for up to several miles. That approach is crucial for shale hydrocarbons, which lie in "thin" horizontal formations and, without horizontal drilling, would not be economical to produce.

An important prior development came from engineers and scientists at national labs (Sandia and others), who developed technologies in partnership with the Gas Research Institute (now the Gas Technology Institute) to peer deep underground. Using microseismic tools and sensors, the engineers were able to

Thus, some see the natural gas boom gas as a cleaner "bridge" to a future with more renewable energy because it produces half the carbon dioxide that burning coal does to generate the same amount of electricity.

"see" the shale deposits and watch how those deposits change with extraction. They could then guide drills directly into underground gas pockets or concentrations of oil. The new technologies dramatically reduced the number of "dry" holes.

Once successfully demonstrated by Mitchell Energy, the combination of underground vision tools, horizontal drilling, and hydraulic fracturing touched off a drilling boom in gas and oil shale formations in other regions of the country, such as the North Dakota's Bakken Shale and the Marcellus Shale under West Virginia, Ohio, and Pennsylvania. The result has been a flood of domestic natural shale gas and shale oil that turned the United States into the world's largest producer of petroleum and gas products in 2014—an astonishing development. With gas supplies abundant and prices low, companies like Dow Chemical have invested billions of dollars in new chemical-manufacturing facilities in the United States, creating jobs and boosting the economy. Utilities are switching from coal to cheaper, cleaner gas, reducing pollution and greenhouse gas emissions. U.S. coal use has dropped 18 percent since 2007. Meanwhile greenhouse gas emissions in the United States have fallen by about 10 percent since 2005—in part because of the substitution of natural gas for coal, but also because of the Great Recession, the growth of renewables, and improved efficiency.

Although an energy boon, the rapid growth of hydraulic fracturing has generated controversies about its safety and environmental and health effects. Drilling for shale gas has sometimes been associated with triggering small earthquakes as well as with contaminated drinking water, polluted streams, and illness. The thousands of gallons of water needed in each well to break open the shale has led to concerns about streams and wells going dry—though engineering improvements make it possible to recycle the fracturing water, and efforts are under way to eliminate the use of water all together. Keeping in mind that since the late 1940s about a million wells have been drilled with a total length of 150,000 miles, the problems are relatively few and appear manageable with innovation and regulation.

Another concern is what the boom in natural gas may mean for the climate. Long term, according to the 2014 National Climate Assessment, if carbon dioxide keeps accumulating in the atmosphere at the current rate, the world could warm by as much as a dangerous 10 degrees Fahrenheit by the end of the century. With moderation of the accumulation rate by significant emission reductions globally, the assessment estimates, the increase could be

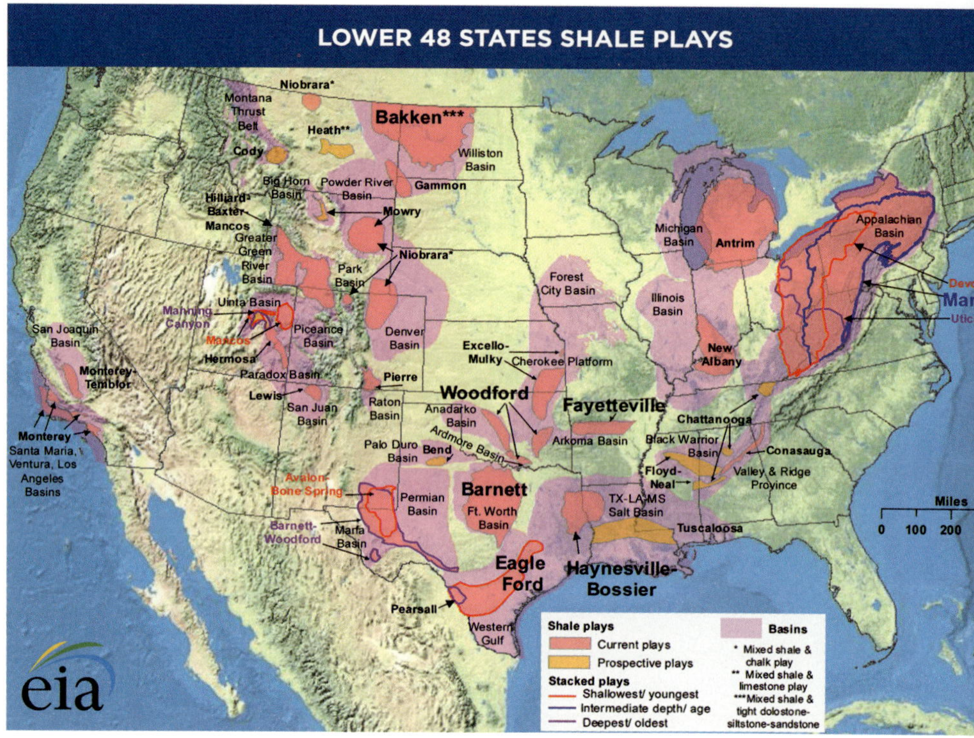

Shale gas is found in shale "plays"—shale formations containing significant accumulations of natural gas. As of 2009, 87 percent of the natural gas consumed in the United States was produced domestically.

as little as 3 degrees Fahrenheit. Thus, some see the natural gas boom as a cleaner "bridge" to a future with more renewable energy because it produces half the carbon dioxide that burning coal does to generate the same amount of electricity. Others worry that, without sufficient control of the extraction process, natural gas leaking from wells and pipelines could put more greenhouse gases into the atmosphere than burning coal. As a practical matter, however, even in the most optimistic scenario, renewable resources will not meet America's energy needs for 30 to 50 years. For that reason, many experts argue that the switch from coal to natural gas is a welcome development.

The good news from the standpoint of the environment is that some of the technology needed to curb greenhouse gas emissions from the use of fossil fuels already exists. Power plant industry engineers have developed and successfully demonstrated processes using chemicals like amines or chilled ammonia that capture the carbon dioxide from smokestacks.

What to do with all that captured carbon remains a challenge. Scientists and engineers are investigating a number of ideas. One key is to pump the carbon deep underground to sequester it from the atmosphere. Another possibility is to use carbon in products like concrete by combining exhaust carbon dioxide, water, and calcium.

Renewable Energy

Even as engineers work to find ways to deal with carbon in the oil, gas, and coal industries, engineering innovations are boosting alternative energy sources. A good example is wind-power technology. Using taller, stronger towers; huge carbon fiber blades more than 250 feet long; better aerodynamics; and improved software and controllers, engineers at companies like Vestas, Siemens, General Electric, and Gamesa have created electric generators powered by wind that are more powerful, more efficient, and more cost-competitive than those in use just a few years ago. In 2012, the United States added more new electric power generation capacity from wind than from any other source, even though the price of natural gas was low. (Of course, turbines produce on average less power than their rated capacity because the wind doesn't always blow.) Now, in 2014, countries and companies can envision a major additional expansion of wind power, as engineers figure out how to safely erect giant wind turbines in coastal waters to tap into powerful offshore breezes, and how to solve the challenges of storing and distributing the energy so that electrical power will be available when wind speeds drop.

The progress in solar energy has been equally dramatic. In 2014 a solar panel costs one-tenth of the price in 1990—and one-hundredth of the price in 1977—due to a whole series of improvements. Engineers have created more efficient processes for making the polycrystalline silicon thin films and other materials used for solar photovoltaic panels. They've improved the efficiency of solar cells so the cells capture more of the sun's energy as electricity, and they've increased the usable yield from the lithographic tools that make the cells. They have also

At the Reese Technology Center in Lubbock, Texas, the DOE/Sandia Scaled Wind Farm Technology (SWiFT) facility's advanced testing and monitoring will help researchers evaluate how larger wind farms can become more productive.

As the cost of solar panels has come down, residential installation has risen. In 2013 in the U.S. solar power was second only to natural gas in new electricity generation.

New York City's iconic Empire State Building has undergone an energy-efficiency upgrade that saved $2.4 million in its first year. The upgrade included refurbishing all 6,514 windows, new building management systems controls, a Web-based tenant energy management system, and elevators that can send excess energy back to the building's grid.

figured out how to install panels more cheaply.

The result has been a rapid acceleration in adoption of solar energy. In 2013, solar outpaced wind in new electricity generation capacity in the United States, coming second behind natural gas. The state of California alone added more rooftop solar systems in 2013 than over the previous 30 years, bringing the state's total solar capacity to 4,000 megawatts—as much as two or three big nuclear plants. Of course, in 2014 solar and wind are still small contributors to the nation's electricity supply, at about 7 percent of overall generating capacity (bringing the total of power from renewable sources, with hydro, to about 15.5 percent of all national electric power requirements).

The alternatives to fossil fuels also go beyond renewable energy. For example, although cheap natural gas has killed plans for some new nuclear power plants and accelerated the retirement of existing ones, two new nuclear units are under construction in Georgia and new technologies for smaller, modular nuclear plants are being considered. Meanwhile, despite daunting economic challenges, technical progress has been made in unlocking the energy stored in plant cellulose, and one factory in Mississippi makes biofuels from feedstock like yellow pine. Perhaps most encouraging and promising, the world continues to make huge strides in energy efficiency, especially in buildings. In fact, U.S. consumption of both electricity and gasoline has declined since 2007, in part because of the recession but also because of improved efficiency.

The key point: the world now has the technological capability to rely on more diverse sources of energy. In some regions, such as Hawaii, where electricity is costly, renewable power already has become economically and operationally competitive. In others, as California's experience demonstrates, policy decisions and incentives can tip the balance toward cleaner, more sustainable sources, while also stimulating further technological innovations.

Looking back to the dark days of past energy crises, who would have predicted then that in 2014 oil and gas would be plentiful in the United States? That massive wind farms would sprout up everywhere from the Texas plains to the seas off the coast of Sweden? And that in 2013 Denmark would produce one-third of its electricity from wind alone—with a goal of 50 percent by 2020? True, we still worry about a sustainable future as the consequences of our warming planet become clearer. But the technological advancements of the last quarter century allow us to be hopeful, if still cautious, about our options.

Humans now live longer and healthier lives than at any other time in history. Average life expectancy for a child born this year in the United States has climbed to 79 years, up from 75 years in 1990 and 70 years in 1964.

A Healthier World

Zapping the heart with little pulses of electricity from implantable pacemakers has prevented millions of deaths, boosted lifespans, and improved quality of life since the late 1950s. And by 2008, the devices had shrunk to less than two inches across.

But that wasn't small enough, thought engineers at medical device maker Medtronic. They set out to build a pacemaker one-tenth that size—a device so small it could be implanted inside the heart by threading it up through a blood vessel. That would eliminate the need to make an incision in the chest to insert the electrical leads for the traditional pacemaker, which sits in a pocket under the skin. Moreover, by fitting inside the heart, the new miniature device, known as Micra, would eliminate the most problematic part of the system—the electrical wire from the device to heart itself.

Was it even possible? The engineers were forced to rethink the whole approach, designing all the electronics—even the battery—as one unit instead of as an assembly of individual components, as in previous devices. They also needed to slash the power consumption dramatically so that a tiny battery could last for 7 to 10 years. They succeeded. In late 2013, the world's smallest pacemaker was implanted in its first human patient. It is now undergoing clinical trials to verify safety and effectiveness.

The tiny leadless pacemaker is just one of the countless medical advances created by engineers over the last quarter century. Doctors

In the course of the last half century, biomedical engineers have reduced the size of pacemakers from external boxes about the size of a car battery to implantable devices barely larger than a 9-volt battery *(left)*. A proposed new model—as tiny as a vitamin capsule *(above)*—could be threaded into the heart through a blood vessel.

After a few sessions of training with a Modular Prosthetic Limb (MPL) developed at Johns Hopkins Applied Physics Lab, Tech Sgt. Joe Delaurier could control the MPL via signals generated by muscles beneath the skin of his residual limb.

can now replace faulty heart valves with a catheter threaded through an artery instead of open-heart surgery. They can prop open narrowed arteries with stents that dissolve, if desired, after doing their job. Improvements in DNA sequencing have helped researchers spot genes linked to Alzheimer's disease and other scourges. The resulting explosion of genetic knowledge, in turn, has led to targeted cancer drugs with fewer side effects and to new ideas for treating other diseases.

Meanwhile, breakthroughs abound in other arenas. The first vaccine that prevents cancer (a vaccine against human papillomavirus, which causes cervical cancer) was introduced in 2006. Prosthetic limbs have enabled amputees to run and dance—and advanced to the point where they can be controlled by electrical impulses from the brain.

In June 2014, researchers at Boston University and Harvard University reported successful results, with an artificial pancreas, which uses sensors, a smartphone, and an insulin-delivery system to precisely control the blood sugar levels in a small sample of people with type 1 diabetes. And neurological diseases like Parkinson's and epilepsy are being treated with electrical stimulation of the brain.

Humans now live longer and healthier lives than at any other time in history. Average life expectancy for a child born this year in the United States has climbed to 79 years, up from 75 years in 1990 and 70 years in 1964. Even greater gains have been made all over the world. Just one technology—vaccination—has eradicated smallpox, virtually eliminated polio, and dramatically reduced measles, mumps, and diphtheria. Child mortality rates continue to drop.

Forging a healthier world is not just a matter of preventive technology like vaccines, however. Great improvements have come from changes in social attitudes and habits as well. For example, decreases in smoking brought tobacco-related deaths in the United States down by about 35 percent between 1987 and 2002. Higher seat belt use and crackdowns on drunk driving, in addition to numerous safety improvements in vehicle design, have cut traffic deaths to just over 10 per 100,000 Americans, down from 16 in 1995 and 23 in 1950.

Some of the current trends are going in the wrong direction. Today, according to United Nations estimates, people with preventable waterborne diseases occupy half of the hospital beds worldwide. In the United States, the increasing incidence of obesity and of diseases triggered or exacerbated by lifestyle, such as diabetes, is threatening to roll back gains in lifespans.

Medical technologies and procedures also can raise the cost of health care without actually improving medical outcomes. And millions of Americans—and billions of people

PRECIOUS COMMODITIES

Safe drinking water is scarce in many parts of the world, where people often have to walk great distances to a source of clean water and then carry heavy containers back to their homes *(above)*. Solving that problem would be an enormous contribution to human health. Meanwhile, bioengineers have created gene-spliced crops, such as soybeans *(below)*, with increased yields to help feed the world's hungry.

around the world—lack even basic health care.

Clearly, good health is about far more than sophisticated MRI machines or other cutting-edge technologies. But engineering has key roles to play in public health challenges. Basic engineering technology can bring proper sanitation and clean water to millions of people who now lack safe water, making perhaps the single greatest possible contribution to human health. Development of electronic health records, interconnected information systems, and data mining techniques can help doctors compare health outcomes after different treatments. The tools of system engineering then make it possible to design and monitor more effective health care delivery processes. Meanwhile, our increasingly interconnected world allows telemedicine and robotic surgery to deliver quality medical care in currently underserved regions, whether in the United States, Africa, or war zones in Afghanistan. And new devices—purchased over the Internet for about $100—can monitor physical activity and diet and might help people lead healthier lives.

As the last half century has demonstrated again and again, people naturally embrace the innovations that improve their lives and offer new capabilities. Engineering has enabled us to leap from tinny-sounding transistor radios and rotary dial phones to smartphones, from bulky black-and-white TVs to giant flat displays and virtual reality. We have altered genes to boost crop yield; developed new materials that make tennis rackets more powerful and airplanes faster, safer, and more fuel efficient; reduced pollution; and developed new sources of energy. Entire industries have been transformed, from publishing and manufacturing to retail and politics.

New engineering creations have enriched our lives, expanded our potential and our reach, even deepened our understanding of what it means to be human and of where we fit into the universe. Nor is this the end of the story. As a peek into universities, national laboratories, and companies around the country would quickly show, the pace of innovation isn't slowing.

The path to the future will never be easy or smooth, of course. The sobering truth is that even as engineering invents ways to solve myriad human and societal problems, the solutions themselves may have unintended, adverse consequences. So it is through a combination of individual choices and public policies that we constantly strive to maintain the right balance of benefit and cost. The heartening truth is that as costs become burdensome, the challenge to restore balance will be met by the most inexhaustible resource that we have—human ingenuity, which gives us discoveries derived from science and innovations created by engineering.

The Next 50 Years

LOOKING TO THE FUTURE

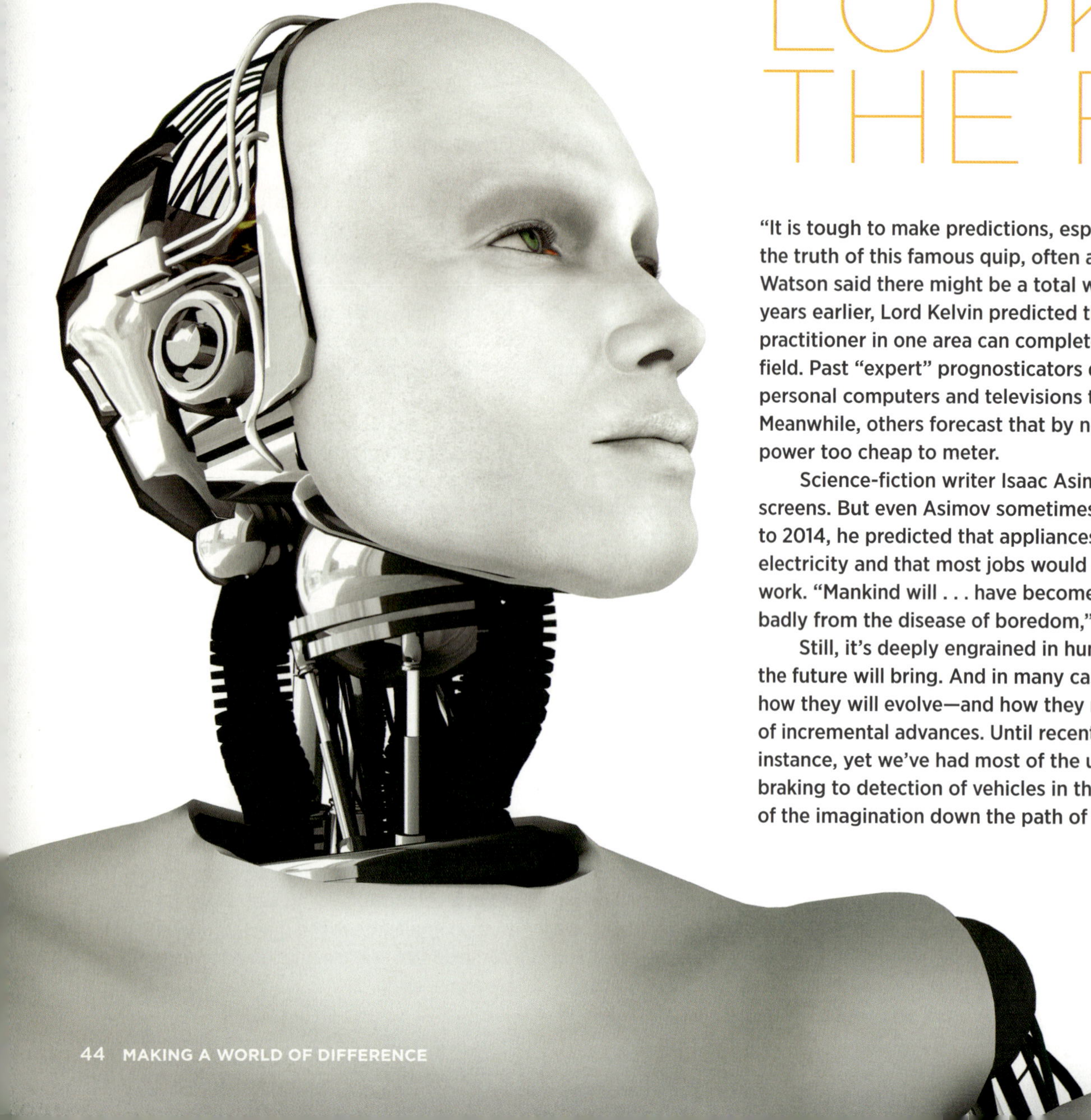

"It is tough to make predictions, especially about the future." Countless examples attest to the truth of this famous quip, often attributed to Yogi Berra. In 1943, IBM chairman Thomas Watson said there might be a total world market "for maybe five computers." Forty-four years earlier, Lord Kelvin predicted that "radio has no future," proving that a brilliant practitioner in one area can completely miss the significance of developments in a different field. Past "expert" prognosticators doubted the utility or appeal of everything from personal computers and televisions to online shopping and overnight package delivery. Meanwhile, others forecast that by now we'd have flying cars, colonies on Mars, and fusion power too cheap to meter.

Science-fiction writer Isaac Asimov correctly anticipated videophones and giant flat TV screens. But even Asimov sometimes got it wrong. In a 1964 essay looking ahead 50 years to 2014, he predicted that appliances would be powered by radioisotopes rather than electricity and that most jobs would be done by machines, freeing up people from actual work. "Mankind will . . . have become largely a race of machine tenders . . . [and] will suffer badly from the disease of boredom," he wrote.

Still, it's deeply engrained in human nature to gaze into a crystal ball and imagine what the future will bring. And in many cases, we can look at today's technologies and anticipate how they will evolve—and how they may bring surprising changes that emerge from a series of incremental advances. Until recently, driverless cars seemed like a distant dream, for instance, yet we've had most of the underlying technologies—from computer-controlled braking to detection of vehicles in the next lane—for years. So it's worth taking a journey of the imagination down the path of continued development of today's technologies.

Brought to You by Engineering

Over the next half century, we can foresee tackling—and solving—many of the pressing problems facing humanity and society today. An NAE report in 2008 describes 14 Grand Challenges for Engineering, such as creating better medicines, restoring and improving our cities, and providing more sustainable sources of energy. Yet even as some of these challenges are met, new issues will arise, sometimes in the form of adverse unintended consequences of our successes. In every case, engineering will be critical to the solution.

Perhaps most important, though, while our imaginations may be spot-on in some cases, in many others the future will be far different than what we now foresee. It will bring answers to questions we aren't asking, and solutions to needs we don't know we have. It will enrich and enhance human lives in ways that are simply impossible to predict—surprising and delighting us, and creating innovations that soon will seem impossible to live without. Mobile phones, for example, were a staple of science fiction and a goal of engineers for years, but the first clunky models were something of a hard sell, and we certainly didn't know we needed smartphones—or social media—until suddenly we did. Today's youth find it hard to believe that previous generations could function without these inventions.

Whatever the shape of the future, the underpinnings and most of the details will come from engineering innovations. As computer scientist Alan Kay, president of the Viewpoints Research Institute once said, "The best way to predict the future is to invent it"—and that's precisely what engineering does.

Grand Challenges

Foremost among the challenges are those that must be met to ensure the future itself.

In 2007 NAE at the request of NSF, convened a diverse international panel of some of the most accomplished engineers and scientists of their generation. The panel's task: to consider broad realms of human concern—sustainability, health, vulnerability, and joy of living—and propose a set of the challenges most in need of 21st-century engineering solutions.

The panel did not attempt to include every important goal for engineering. Rather, it chose the problems we must solve to ensure survival of a livable Earth and the well-being of its inhabitants. Earth's resources are finite, and our growing population currently consumes them at a rate that cannot be sustained. Among the most pressing concerns, then, is the need to develop new sources of energy while also preventing or reversing the degradation of the environment. Another is to find new methods to protect people against pandemic diseases, terrorist violence, and natural disasters. The engineering solutions to challenges such as these can no longer be designed solely for isolated locales, but must address Earth as a whole and all the planet's people. As the panel concluded in its 2008 report, "a world divided by wealth and poverty, health and sickness, food and hunger, cannot long remain a stable place for civilization to thrive."

- **Make solar energy economical**
- **Provide energy from fusion**
- **Develop carbon sequestration methods**
- **Manage the nitrogen cycle**
- **Provide access to clean water**
- **Restore and improve urban infrastructure**
- **Advance health informatics**
- **Engineer better medicines**
- **Reverse-engineer the brain**
- **Prevent nuclear terror**
- **Secure cyberspace**
- **Enhance virtual reality**
- **Advance personalized learning**
- **Engineer the tools of scientific discovery**

To learn about the Grand Challenges for Engineering visit the project's interactive website at **www.engineeringchallenges.org**

New Materials, New Possibilities

From the Stone Age to the Iron Age, epochs of human history have been named after materials. That's not surprising, because new materials open the door to entirely new and unexpected applications and developments, weaving new threads into the tapestry of human progress and changing how we live and work. Today's Information Age might justly be called the "Silicon Age" because of the enormous capabilities provided by silicon-based devices and applications, although modern advances have also required numerous other crucial materials, from optical fibers to high-strength alloys. The silicon frontier will be extended farther, no doubt, but new materials may take us beyond a simple extrapolation of today's technology, meeting the needs that we don't yet know we have.

A new material—graphene—is a layer of carbon only one atom thick, discovered by physicists Andre Geim *(bottom, left)* and Konstantin Novoselov, who won the 2010 Nobel Prize in Physics for their groundbreaking experiments.

Scientists and engineers are now hard at work creating and exploring the potential of new materials. In the early 2000s, for example, physicists Andre Geim and Konstantin Novoselov at the University of Manchester in England were tinkering with graphite and tape. They realized that it was possible to peel off a layer of carbon so thin that it was only one atom thick. This material, dubbed graphene, was almost completely transparent, yet so dense not even helium could pass through. Possessed of immense strength, it also had interesting electrical properties, which Geim and Novoselov nailed down by studying dozens of ultrathin electronic devices they made from graphene. Their work won them the 2010 Nobel Prize in Physics—and pointed to a new path for devices. "Graphene could change the electronics industry, ushering in flexible devices, supercharged quantum computers, electronic clothing and computers that can interface with the cells in your body," predicted the *New York Times* in 2014.

Meanwhile, in Hewlett Packard's Quantum Science Research Lab, Stanley Williams has built novel devices with a completely different approach. His idea: Use chemical reactions to grow switches and wires that assemble themselves into circuits. Working with colleagues at UCLA, Williams fashioned the world's first molecular logic gate, the building block of digital circuits. If such "molecular electronics" devices could be used to create viable computers, they could put the power of a hundred workstations on a chip the size of a grain of sand.

We don't know if the central processors of computers 25 or 50 years from now will be built from graphene, self-assembled molecules, DNA, or any of a number of other exotic materials emerging from today's laboratories. We do know, however, that the enormous advances in materials (and concomitant leaps in computing power) that have already transformed our lives in 2014 will continue—perhaps even accelerate. Engineers will create ever-smaller devices, exploiting the strange world of quantum mechanics, where atoms can exist in different places at once and affect each other across considerable distances. "Materials with genuine quantum properties will have enormous impact," says Venkatesh Narayanamurti, professor of technology and public policy at Harvard University's School of Engineering and Applied Sciences. And other improved or new materials will enable continual advances in everything from cars and planes to the buildings we live in.

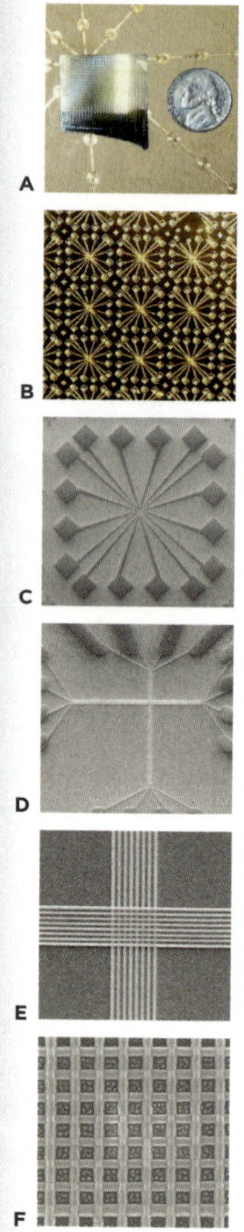

In this series from HP Labs, each successive image is magnified about 10 times the previous one, from (A), the wafer on which 625 64-bit memories are imprinted, through (F) a close-up of a single memory, with one bit stored at each of the 64 intersections.

The future is now: Flying drones that deliver packages and cars that drive themselves are already being tested, while virtual reality software helps train aviators both to fly a plane and to jump out of it.

A World of Embedded Intelligence

In 2014, we already have sports watches that record workouts and autonomous flying drones the size of birds. But imagine dramatically shrinking those devices and many others, while also adding the raw computing power of today's supercomputers. Imagine similar giant leaps in sensors, communications capabilities, displays, software, batteries, and mechanical actuators. Put all those together and we can embed intelligence in virtually anything—from light bulbs and refrigerators to cars and complex manufacturing tools. Already, smart devices can answer simple inquiries and understand simple commands. It's not a stretch to predict that these capabilities will improve enough to make it appear that devices are thinking, speaking, and acting independently. Some of these devices will fail in the market, but others will hit the sweet spot that delights consumers and improves or enhances their lives.

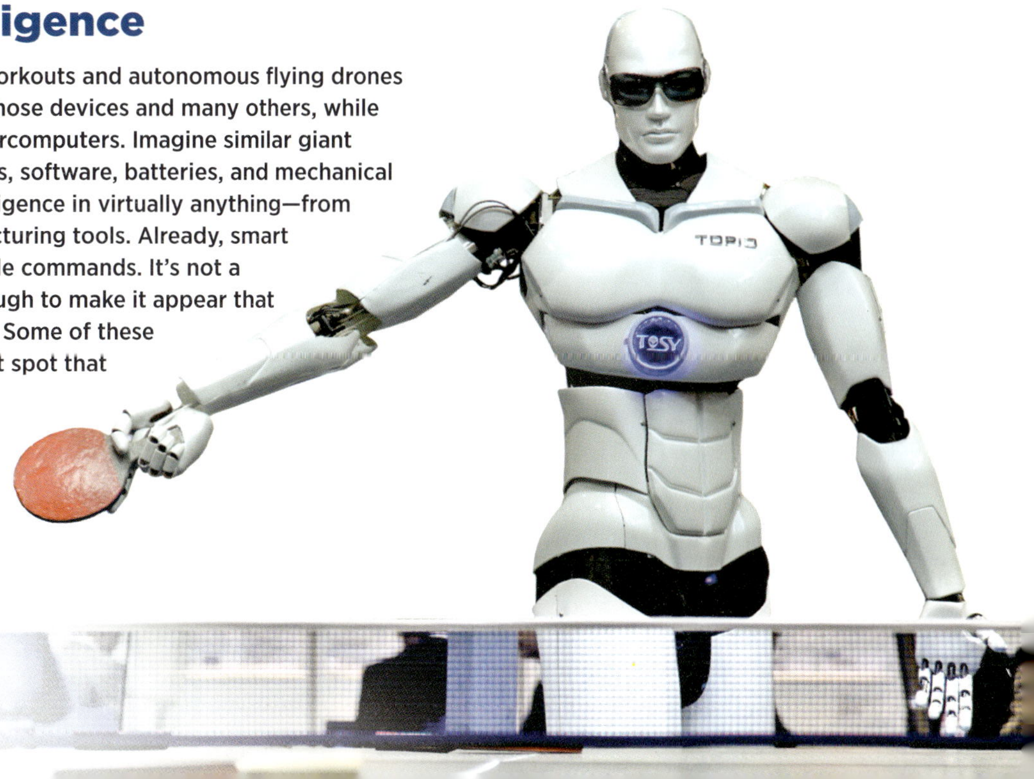

Here's just a sample of what may be possible, some of which is already taking shape: Virtual reality technology that trains the military. Cars that drive themselves, in constant communication with other vehicles and with traffic signals. Appliances and houses that respond to voice commands—maybe even know what you want automatically. Displays that cover entire walls, enabling us to put an art gallery with treasures like the *Girl with a Pearl Earring* in our homes, visit with grandma in what feels like an adjacent room, negotiate a deal across the "table" with partners in Tokyo or Kazakhstan, have a prime seat at the opera or rock concert, work with a personal trainer, or take a virtual climb up Mount Kilimanjaro. One of the NAE's Grand Challenges, enhancing virtual reality, will be easily met, predicts Ray Kurzweil, now tackling natural language understanding at Google. "By the early 2020s we will be routinely working and playing with each other in full immersion visual-auditory virtual environments," he writes. Another Grand Challenge, tailoring education to meet individual needs, will also be met, says Leah H. Jamieson, dean of engineering at Purdue University. "I absolutely believe it will be possible to build interactive systems that provide personalized learning environments."

The future world could bring what Asimov anticipated a bit too early—the creation of robots that read, learn, and even feel. Such robots could take care of the elderly, file tax returns, build houses, and discuss the origins of the universe or the latest escapade of the next generation of reality TV stars. "We will have another intelligent species on Earth," predicts Danny Hillis, chairman and cofounder of Applied Minds, LLC.

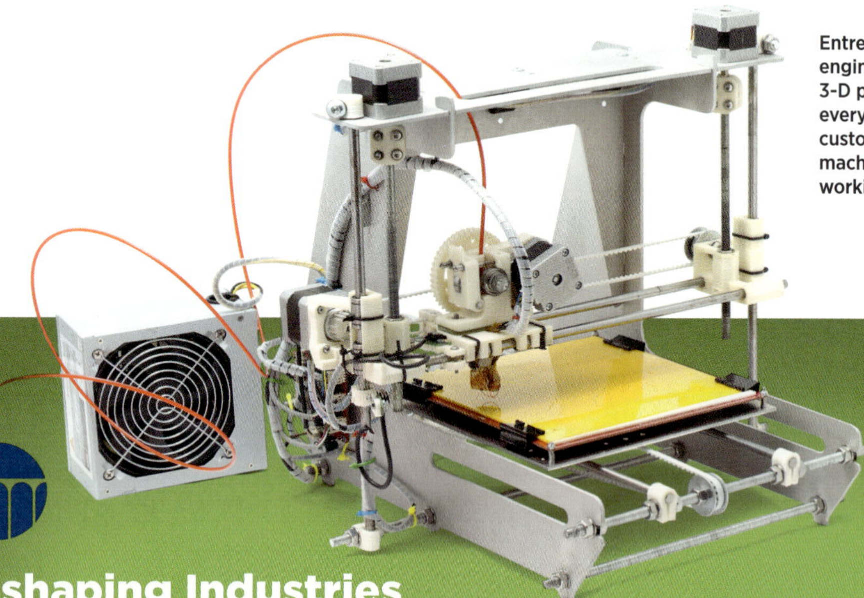

Entrepreneurs and engineers are using 3-D printers to create everything from custom toys and machine parts to working prostheses.

Reshaping Industries

New materials are one driver of change. Development of new manufacturing methods and tools is also crucial—another job for engineers. Consider 3-D printing. GE Aviation used to make jet engine nozzles by welding together 18 different parts. But not for its latest, most efficient engine. The company now builds the nozzles one layer at a time by precisely depositing material with a 3-D printer, in much the same way an ink-jet printer sprays on paper.

3-D printing is a potential game changer for today's factories, warehouses, supply chains, distribution systems, and delivery companies. It also has the potential to eliminate the waste of raw materials in manufacturing processes. Instead of machining or forging a part like a connecting rod, a 3-D printer puts material just where it's needed, like an oyster building up its shell layer by layer. "3-D printing sounds trite, but you can build structures that you could never do any other way," says Paul Citron, retired vice president for technology policy at medical device maker Medtronic. This technology makes it possible for anyone to become a manufacturer. "Imagine that instead of having to stock parts at an auto supply store, the guy goes to a keyboard when you ask for a part. He then makes the part at his 3-D printer and hands it to you," says John Wall, vice president and chief technical officer at engine and power systems manufacturer Cummins. Or imagine inventors dropping by the local 3-D print shop to print out working prototypes of their latest ideas. You could even print stuff in your own home.

It's also theoretically possible, if you have the right materials, to print almost anything, including living tissue. If you needed a new liver, say, doctors might extract a few of your stem cells, transform them into liver cells and print out your new organ. "By the early 2020s we will print out a significant fraction of the products we use, including clothing as well as replacement organs," predicts Google's Kurzweil.

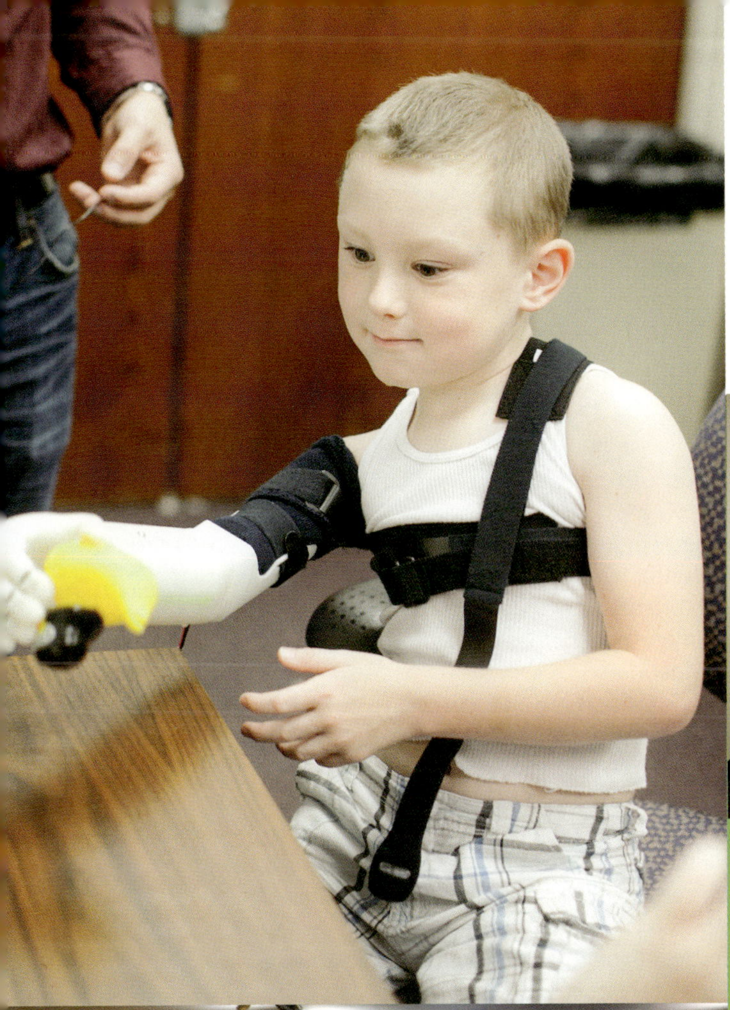

Born without an arm, six-year-old Alex Pring of Groveland, Florida, practices picking up objects with his new 3-D printed prosthetic arm and hand, designed and made by engineering students at the University of Central Florida for about $350.

GE Aviation's new jet engine *(left)* includes a fuel nozzle made by 3-D printer.

Expanding Our Connections

Many visionaries foresee that people in future decades will want to be connected even more than they are today, and that such connections will improve their quality of life. If so, engineers will be the architects of this hyperconnected future. "The connectivity of everything is within a decade," predicts Charles Holliday, Jr., former CEO of du Pont. "It will change how we think about managing our lives." And by 2025, "information sharing over the Internet will be so effortlessly interwoven into daily life that it will become invisible, flowing like electricity, often through machine intermediaries," according to a 2014 report from the Pew Research Center's Internet Project. The developing world will continue to leapfrog the old wired infrastructure, as remote villages connect to the larger world with wireless broadband networks.

As is frequently the case with new technologies, hyperconnectivity will offer challenges along with opportunities. Will the regulations written for telephone communications need to be rewritten for the Broadband Age? Can cybersecurity efforts not only keep the hackers at bay but also keep criminals and terrorists in check? Can we find a balance between hyperconnection and personal privacy that is acceptable to most people? Governments and societies will need to grapple with these questions and challenges, but engineering advances will underpin the solutions.

Making Energy Sustainable

Intelligent, hyperconnected devices, 3-D printers, and other technologies will bring surprises, meet unanticipated needs, and change our lives in ways that are hard to imagine. But some aspects of the future are easier to predict. To create a better, richer, and healthier future for all people and nations, we know we must tackle and solve problems that are already obvious now.

J. Craig Venter *(far left)* is working on synthesizing algae to replace fossil fuels. In France an experimental thermonuclear reactor project is under construction *(above)*.

One of those big challenges is creating a sustainable supply of energy. Energy is crucial to maintaining and boosting standards of living. To bring billions of people out of poverty, therefore, we'll either need more energy or huge improvements in energy efficiency—or, most likely, both. But right now, because of our dependence on fossil fuels, humans are emitting carbon dioxide and other greenhouse gases into the atmosphere at a rate that exceeds anything the Earth has experienced in millions of years. Since 1900, the planet has warmed by about 0.8 degree Celsius (1.5 degrees Fahrenheit), and the number of extreme weather events scientists are linking to warming has increased in recent years. There's "compelling evidence that increasing temperatures are affecting both ecosystems and human society," warns the 2014 National Climate Assessment.

Thus, the energy mix is as important as—or more important than—the total energy needed. If we want to avoid contributing to the carbon dioxide buildup by burning fossil fuels, efforts both to switch to renewable or other low-carbon power and to use less energy must go forward. Wind power and solar power currently represent about 7 percent of overall generating capacity, and engineering advances there and elsewhere are in the offing. Improvements in wind turbines and solar panels, for example, are rapidly making them more efficient and cheaper, and new battery technologies promise to solve the problem of intermittency. Argonne National Laboratory, for instance, is leading a major multi-institution effort to build a battery with five times the energy density of today's best, at one-fifth the cost. Such batteries could also make electric cars far more practical and attractive, weaning much of the transportation sector from the fossil fuel pump.

Huge improvements are possible in using energy more efficiently. Something as simple as better insulation, such as ultralight aerogels, can dramatically reduce the energy needed to heat and cool homes and factories and run refrigerators.

Engineers are also working to design safer, cheaper nuclear reactors. As a virtually carbon-free source of reliable energy, "nuclear power has to play a significant role in the future," says Cummins's John Wall. It may also be possible to harness the fusion reaction that powers the sun—another NAE Grand Challenge. A research reactor, the International Thermonuclear Experimental Reactor Project, is now under construction in Cadarache, France. Although fusion energy still faces daunting technical hurdles, many experts remain hopeful. "I think we'll have fusion, maybe not in 50 years, but eventually," says Julia Phillips, vice president and chief technology officer at Sandia National Laboratories.

Meanwhile, other creative ideas abound. For example, Caltech's Frances Arnold, winner of the 2011 Draper Prize, is using the techniques of directed evolution to produce new biocatalysts to convert cellulose to sugars and then to biofuels. In other labs researchers use catalysts and other materials to mimic photosynthesis and capture energy from sunlight. At least five different designs are competing to turn the energy from ocean waves or tides into electricity. Smart micro-grids promise not only to keep the lights on in U.S. cities, but also to bring renewable power to remote villages in developing countries, bypassing the need for expensive power lines and central power plants.

Some visionaries believe human ingenuity and engineering wizardry can easily wean humanity from fossil fuels within 50 years. Kurzweil, for one, predicts that "by 2030 solar energy will have the capacity to meet all of our energy needs"—including providing enough extra power to purify vast amounts of salty water. Meeting the Grand Challenge of making solar energy economical thus could also satisfy the growing need for clean water, another Grand Challenge. A surplus of energy would also make it possible to power scrubbers that can pull carbon dioxide and all other forms of pollution from the air, says Cherry Murray, dean of Harvard University's School of Engineering and Applied Sciences.

The conventional wisdom, though, is that wind and solar alone can't provide enough energy for a growing world, especially when the wind dies or the sun sets. Many experts insist that the world will depend on fossil fuels for a sizeable percentage of its energy for a least 50 more years. "Energy is going to come from a lot of different sources," says Holliday. In particular, if for no other reason than it is plentiful and cheap, the world is unlikely to stop burning coal soon, with 2,300 existing coal plants and more than 1,000 proposed new facilities. So, to reduce emissions in the medium term, even for the long-term, wide-scale implementation of improved technologies for grabbing the carbon from fuel or carbon dioxide from smokestacks is essential. And additional innovations are needed to pave the way for safe storage of that carbon, meeting the Grand Challenge of developing carbon sequestration methods.

Creating a cleaner, more sustainable energy future will require hard decisions based on data and evidence, which can come from engineering advances such as more powerful supercomputers and sophisticated sensors on land, in the oceans, and in space. The decisions themselves are typically outside the realm of engineers and scientists—but scientists and engineers will need to engage them as they work to create solutions to the world's energy problems. As a practical matter, according to the National Climate Assessment and a joint report by the U.S. National Academy of Sciences and the Royal Society in the United Kingdom, carbon dioxide concentrations and global temperatures presently in place make some climate impacts inevitable, even if greenhouse gas emissions were to cease. So, as we hedge our bets by striving to change the energy mix, engineers also face the challenge of helping society adapt to the changing global environment.

Feeding the World's Billions

The combination of massive harvesters and other farm machinery, precision fertilization, genetically modified crops, and other advances has dramatically boosted yields of corn and other crops across the United States and around the world. At the dawn of the 20th century, about 50 percent of the U.S. population was involved in food production. Today that number has dropped to 2 percent.

But to feed the world's growing number of people, we will need another increase in productivity. One of many possible answers to this problem is harnessing the potential of genetic engineering and fermentation. Biological engineers are already growing gene-spliced algae that make a full set of the protein building blocks, or amino acids, that we need in our diet. Turn that algae into flour, and we could replace millions of acres of amber waves

"We'll have local food in home gardens, hanging gardens, and hydroponic gardens in all sorts of interesting places."

of grain with giant stainless steel vats filled with fermenting microbes. Slip in the genes for muscle and blood proteins like actin and myogloblin, along with genes for healthful fats, and algae or other microbes could even make what J. Craig Venter, chairman and president of the J. Craig Venter Institute, dubs "motherless meat," ending the need for a home on the range.

Venter calculates that microbial factories could produce as much food as our current system of agriculture using only one-tenth the land area. If we wanted, we could turn the Great Plains back into a vast prairie teeming with buffalo, or bring forests back to many areas of the world that were cleared for cultivation. Plus, the approach would solve one of NAE's Grand Challenge problems—managing the nitrogen cycle to reduce the nutrient pollution that's harming the world's creeks, rivers, lakes, and coastal areas.

Of course, that's just one possibility. Harvard's Murray and others have different ideas for feeding the world's billions of people. Murray predicts that a global disaster—such as a disease that wipes out all wheat or rice crops—will bring a dramatic shift from today's industrial monoculture agriculture to a distributed, local system, where a wide range of plants are grown on rooftops and other spaces throughout cities and communities. "We'll have local food in home gardens, hanging gardens, and hydroponic gardens in all sorts of interesting places," she says. In fact, this trend toward more local food is already beginning—even without a major crop failure.

A Healthier Future

Among the many medical advances from the lab of MIT chemical and biomedical engineer Robert Langer are polymers designed to dissolve at different rates in the bloodstream. Encapsulate a drug or a vaccine inside tiny spheres made from these materials, inject them into the blood, and the microspheres will "deliver" the actual medicines to the site of cancers or other tumors days or weeks later. "It may sound trivial, but it can help change the face of medicine," says Langer.

To fight deadly diseases such as tuberculosis or Ebola, doctors must treat people with multiple, periodic doses of drugs or vaccines. Yet in many parts of the world, it's hard enough to get patients to health clinics once, let alone every few weeks. Microsphere technology solves that problem. Patients could be given full courses of treatment or vaccinations with a single injection. Suddenly, once-intractable diseases can be cured or prevented.

Over the next few decades, bioengineers are expected to create many more such weapons in the fight against infectious diseases. These innovations might include malaria and tuberculosis vaccines as well as cheap, effective (and simple to administer) drugs against HIV, and robust technologies for delivering clean water and providing basic sanitation in underdeveloped countries. By 1980 the world had eradicated smallpox. It's not a great leap of imagination to think that we can finish the job of eliminating polio and make dramatic inroads against cholera, AIDS, diphtheria, and other terrible infectious diseases. The benefits would be enormous, not just in reducing infant mortality and increasing life expectancy, but also in boosting productivity, economic growth, and standards of living all over the world.

But that's just the beginning of how science and engineering have the potential to transform health. Drop in, for instance, at the Stanford lab of Karl Deisseroth, which recently tackled a project so

risky that Deisseroth enlisted only those colleagues whose careers were sufficiently established that they would not be set back by a failure. The idea: make an intact, transparent brain with all of its internal structure and wiring visible. The team succeeded, figuring out how to support a mouse brain with an external hydrogel skeleton, then dissolving away its opaque fat. As a result, researchers can now chart all the connections between neurons, a significant step on the journey toward meeting another Grand Challenge, reverse engineering the brain.

Eventually, with better understanding of brain chemistry as well as brain circuitry and the underlying mechanisms of biology and disease, medical professionals may be able to intervene successfully in everything from addiction and epilepsy to schizophrenia and Parkinson's disease. Meanwhile, researchers predict that advances in understanding the biology of the rest of the body will make it possible to tame autoimmune diseases and cancer.

Similar gains will come from reading humanity's genetic code, from cataloging all of our proteins, and from manipulating genes and biology. Danny Hillis of Applied Minds foresees making real-time measurements of the chemicals coursing through the body, and then using computing tools like data mining and pattern recognition to spot chemical signals going awry—long before any actual symptoms of illnesses appear. "We'll be able to see a problem coming and intervene on the side of the body before we ever get sick," he predicts. Paul Citron, retired from Medtronic, expects that for diabetics "an artificial pancreas will become a reality," staving off the many complications of diabetes by precisely controlling blood sugar. MIT's Langer—who was awarded the 2002 Draper Prize for "bioengineering of revolutionary medical drug delivery systems"—predicts that it will be possible to regenerate spinal cords, to replace failing organs and body parts with engineered tissue and to turn the body into its own drug factory by injecting the manufacturing instructions in the form of messenger RNA. "The combination of biology and engineering will lead to all kinds of new things, improving the quality of care and quality of life," he says.

Just as with computer power and the connected world, these advances could be of tremendous benefit to humanity. We'll get longer, healthier, more productive lives—and, with advances in brain science, a deeper understanding of what it means to be human.

But the technology will also raise difficult questions and ethical dilemmas. Will society be willing (and able) to pay for expensive new treatments and approaches for everyone, or will these advances benefit only the rich? Once it becomes possible, will we rush to tinker with our genes to create new generations with superior athletic abilities or intelligence? It could indeed be a brave new world. "Human engineering will be inevitable," says J. Craig Venter. Once again, how tomorrow's society decides to use its new engineering powers will be crucial.

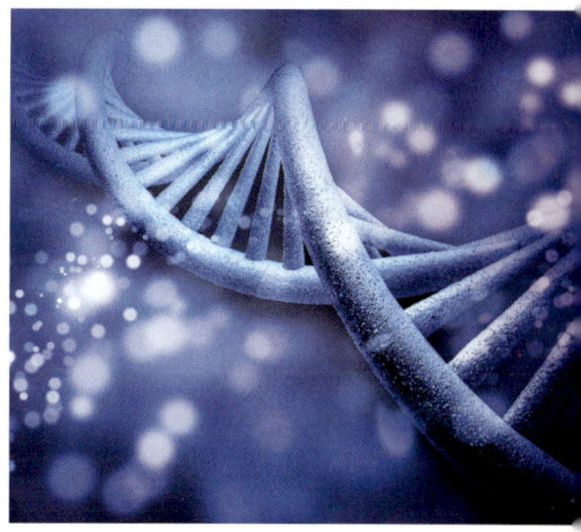

China alone must build the equivalent of a city the size of Boston every 17 days to accommodate the 14 million additional people per year projected to live in the country's urban areas.

Cities, Limits, and New Frontiers

A half century from now, one of the most critical factors determining what the future looks like will be this: how many people will be packed onto the planet? The United Nations' best estimate is that the global population will climb from today's 7.2 billion to 9.6 billion in 2050. But higher fertility could send that soaring past 15 billion by the end of the century. Or if the developing world emulates the low birth rates of countries like Italy and Japan, the number could actually decline by then to 7 billion.

Either way, the consequences will be profound. A more populated world increases the challenges of providing food, health care, and housing—even bumping up against the limits of what the planet can support. On the other hand, lower population numbers mean that average age will climb quickly, making it harder to care for the elderly. The number of people older than 65 is on track to exceed those younger than 15 in most countries within a decade or two—for the first time in human history.

One trend that's safe to predict, however, is increasing urbanization. More than half the world's people now live in cites. A million more are born there or move in every week. China alone must build the equivalent of a city the size of Boston every 17 days to accommodate the 14 million additional people per year projected to live in the country's urban areas. How can we keep all these people from ending up in sprawling shantytowns all over the world?

Many urban planners suggest that the answer lies in taller, denser cities. According to Antony Wood, executive director of the Council on Tall Buildings and Urban Habitat, engineers already know how to build soaring structures two or three kilometers tall. What's harder is maintaining the vitality of urban life in a city of super-skyscrapers. So much of the vibrancy of a city goes on at ground level—in parks, shops, and restaurants. The answer may be to bring that vitality upward. "If a city gets ten times more vertical and ten times denser, then we need to replicate the ground level in the sky—creating urban habitats in the sky," says Wood. That shift would be a major undertaking for urban planners and civil engineers.

Of course, future cities won't be able to function without other vital engineering advances: replacing and redesigning aging water mains and sewers; reshaping transportation systems to cope with higher populations. In the United States, urban engineers can envision a future where the number of cars drops and an increasing proportion are shared. When not in use in the denser, future city, many cars might sit around in automated multistory garages. Need a car? Call one with your smartphone. It may even drive to you and chauffeur you around. When you're done, "push a button and the car parks itself in the parking garage," says Holliday. Urban planners and engineers are already exploring many of these possibilities.

With more than 40 percent of the world's population living within 60 miles of coastlines—and many more along rivers—engineers must also figure out how to make cities more resilient against rising seas, river floods, and extreme weather events. In the aftermath of 2012's Superstorm Sandy, for instance, New York City has developed a detailed plan for reducing the damage from future storms. The engineering steps that New York and other cities could take are as simple as elevating homes and moving the mechanical guts of buildings from the basement to higher floors, or as complex as re-creating and reengineering the buffer of coastal wetlands that can protect cities from raging storm surges.

But engineering the path to the future is not just about planning for disaster, coping with potential limits, or finding solutions to innumerable problems. As this chapter tries to convey, it's also about eradicating diseases; lifting millions out of poverty and sickness; forging stronger, more resilient communities; and preparing for many possible futures. It's about making life richer and more fulfilling. It's about pushing back the frontiers of knowledge—even freeing us from the bounds of Earth. "Within 25 years, we'll go to space as routinely as we go the grocery store today," predicts Wanda Austin, president and CEO of The Aerospace Corporation. "It could be for fun or because it's critical for our survival."

Just imagine what it would mean to use engineering advances to finally understand the mysterious dark matter that makes up most of the universe or to discover extraterrestrial life. "Contributions from engineering will bring many more astonishing insights about ourselves, our Earth, and our universe in the next 50 years," says Princeton University's Robert Socolow.

Decade by decade, century by century, engineering has taken us further and further from the first glimmerings of human art and culture on the walls of Paleolithic caves. And to some experts, it's even helping us leave behind some of the darker side of human nature. Harvard University cognitive scientist Steven Pinker argues that as human society becomes more modern (in large part from technological advances), we become a kinder, gentler species. "You can see [the decline of violence] over millennia, over centuries, over decades, and over years," he says. "We are probably living in the most peaceful time in our species' existence."

It's a highly controversial idea, but a hopeful and attractive one. If the march toward greater enlightenment continues—and the flowers of engineering bloom as they have throughout history—then the next half century really will be worth looking forward to.

VIEW FROM AFAR

A First Step to Other Planetary Systems

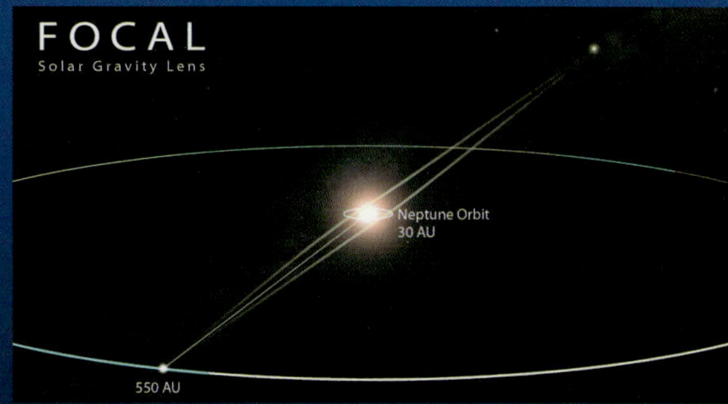

The Kepler Space Telescope, launched in 2009 to search for planets orbiting other stars, has found many such systems, including at least one, Kepler-186, with a planet similar to Earth in what astronomers call the "habitable zone"—the distance from a star at which liquid water can exist. Kepler-186 is 500 light-years away, meaning that light from its star takes 500 years to reach us. With the technology of the next decade or two, 500 light-years is much farther than we can send a probe to do a flyby.

But we can undertake missions in the next few decades that could let us look more closely at some of the planets Kepler has found. Scientists and engineers have proposed to do this by taking advantage of an effect first predicted by Albert Einstein in 1936—namely, that the gravity of large objects would bend light, just as a glass lens does in a traditional telescope. Astronomers already use gravitational lensing to get better images, from our perspective on Earth, of objects located beyond large stars or galaxies. However, they can't "aim" a galaxy or re-position Earth to choose what to examine.

We could use our Sun itself as a gravitational lens—except we can't do it from Earth, or even from Earth orbit; we're much too close. Instead, we have to send a telescope out along the focal axis of the solar lens—to the closest point where the light from the object we want to look at, bending around the Sun, comes into focus *(above)*. The trick is getting our telescope out there. The focus of the solar lens begins 3.2 light-days from the Sun and continues outward, with the image quality improving as the telescope gets farther away from the Sun.

Over the next half century, engineers will develop new probes that will be smaller, lower mass, and easier to propel to high speeds than anything we've launched so far. They will be powered by new propulsion systems, such as ion rockets or "light sails" *(below left)* that catch the solar wind speeding from the Sun at more than million miles per hour. With a push from a laser beamed from Earth, our telescope's light sail could reach the focus of the solar lens in a few years. As engineering advances make probes still smaller and less expensive, we could even launch a swarm of space telescopes to different focal points of the Sun's gravitational lens, giving us close-ups of more distant stars and the means to detect radio or optical signals that might indicate an advanced civilization.

"The new frontier of the 20th century was our solar system" says David Messerschmitt, Roger Strauch Professor Emeritus of Electrical Engineering and Computer Sciences, University of California, Berkeley. "And the new frontier for the 21st century will be interstellar space in our region of the Milky Way galaxy."

REFERENCES

1964: Dawn of the Digital Age

Introduction

George Constable and Bob Somerville, *A Century of Innovation: Twenty Engineering Achievements That Transformed Our Lives* (Washington, DC: National Academy of Engineering and Joseph Henry Press, 2003), 10.

Electronics: Smaller, Faster, Cheaper

"ENIAC: Celebrating Penn Engineering History," Penn Engineering, University of Pennsylvania, http://www.seas.upenn.edu/about-seas/eniac/operation.php.

Time-Life Books, *Computer Basics* (Alexandria, VA: 1985), 59–60.

"Programming the ENIAC," Columbia University Computing History, http://www.columbia.edu/cu/computinghistory/eniac.html 2/6.

"John Bardeen, William Shockley,Walter Brattain: Invention of the Transistor–Bell Laboratories," American Physical Society, 2008, http://www.aps.org/programs/outreach/history/historicsites/transistor.cfm.

"The Nobel Prize in Physics 1956," Nobel Media AB 2013, http://www.nobelprize.org/nobel_prizes/physics/laureates/1956/.

"CDC 6600's Five-Year Reign," Computer History Museum, http://www.computerhistory.org/revolution/supercomputers/10/33.

"The History of Supercomputers," ExtremeTech, April 10, 2012, http://www.extremetech.com/extreme/125271-the-history-of-supercomputers.

"The Genius: Meet Seymour Cray, Father of the Supercomputer," Businessweek, 1990 (updated 1997), http://www.businessweek.com/1989-94/pre88/b31571.htm.

"Heavyweights of the Supercomputing World," TechRadar, February 1, 2009, http://www.techradar.com/news/computing/heavyweights-of-the-supercomputing-world-515223.

Constable and Somerville, *A Century of Innovation*, 54–57.

Paul E. Ceruzzi, *Computing: A Concise History* (Cambridge, MA: MIT Press, 2012), 86–90.

NAE Awards, Draper Prize 1989, http://www.nae.edu/Projects/Awards/DraperPrize/DraperWinners/page19898887/55087.aspx.

"Robert Noyce: Statesman of Silicon Valley," History, Intel Museum, http://www.intel.com/content/www/us/en/history/museum-robert-noyce.html.

"The Nobel Prize in Physics 2000," Nobel Media AB 2014, http://www.nobelprize.org/nobel_prizes/physics/laureates/2000/.

"The Integrated Circuit: 1961–Present," Chips Etc., http://www.chipsetc.com/the-integrated-circuit.html.

Brian Bergstein, "Fortran Creator John Backus Dies," Associated Press, March 20, 2007, http://www.nbcnews.com/id/17704662/#.U46i65RdUY0.

"Scientific Computing's Future: Can Any Coding Language Top a 1950s Behemoth?" Ars Technica, May 7, 2014, http://arstechnica.com/science/2014/05/scientific-computings-future-can-any-coding-language-top-a-1950s-behemoth/.

"Grace Murray Hopper," http://www.cs.yale.edu/homes/tap/Files/hopper-story.html.

"COBOL: Introduction," National Museum of American History, http://americanhistory.si.edu/cobol/introduction.

Time-Life Books, *Software* (Alexandria, VA: 1990), 27–28.

"The Digital Equipment Corporation PDP-8, Frequently Asked Questions," The PDP-8 Collection by Douglas W. Jones, The University of Iowa, Department of Computer Science, http://homepage.cs.uiowa.edu/-jones/pdp8/faqs/.

"PDP-1 Restoration Project," Computer History Museum, http://pdp-1.computerhistory.org/pdp-1/index.php?f=theme&s=1.

"DEC's Blockbuster: The PDP-8," Computer History Museum, http://www.computerhistory.org/revolution/minicomputers/11/331.

"DEC PDP-11: Benchmark for the First 16/32-Bit Generation," CPUShack.com, http://www.cpushack.com/CPU/cpu3.html.

"PDP-7 Definition," Linux Information Project (20005–2007), http://www.linfo.org/pdp-7.html.

Digital at Work: Snapshots from the First Thirty-Five Years, ed. Jamie Parker Pearson (Burlington, MA: Digital Press, 1992), http://www.soemtron.org/downloads/decinfo/decdigitalatwork1992-102630350.pdf.

Harnessing Light: The Power of Lasers and Fiber Optics

Constable and Somerville, *A Century of Innovation,* 206–13.

"Einstein Predicts Stimulated Emission," This Month in Physics History, American Physical Society, http://www.aps.org/publications/apsnews/200508/history.cfm.

Charles H. Townes, "The First Laser," in *A Century of Nature: Twenty-One Discoveries that Changed Science and the World,* ed. Laura Garwin and Tim Lincoln (Chicago: University of Chicago Press, 2003), http://www.press.uchicago.edu/Misc/Chicago/284158_townes.html.

"The Emergence of the Compact Disc," *Communications Magazine,* IEEE, no. 48 (January 2010), 10–17, http://ieeexplore.ieee.org/xpl/login.jsp?tp=&arnumber=5394021.

Lasers in Ophthalmology: Basic, Diagnostic, and Surgical Aspects: A Review, ed. Franz Fankhauser and Sylwia Kwasniewska (The Hague: Kugler Publications, 2003), 115.

"A History of the Laser," National Laser Company, http://www.national-laser.com/laser-history.htm.

"Charles H. Townes–Biographical," Nobel Media AB 2014, http://www.nobelprize.org/nobel_prizes/physics/laureates/1964/townes-bio.html.

http://www.laserfest.org/lasers/history/early.cfm.

"A History of the Laser," National Laser Company.

"Argon Laser Photocoagulation in the Management of Retinal Detachment and Predisposing Lesions," *Lasers in Surgery and Medicine* (1984), http://www.ncbi.nlm.nih.gov/pubmed/6542963.

The World Above and Beyond: Space Exploration

Constable and Somerville, *A Century of Innovation,* 134–41.

Paul Dickson, "Sputnik's Impact on America," excerpted from *Sputnik: The Shock of the Century* (Walker and Company: 2001), http://www.pbs.org/wgbh/nova/space/sputnik-impact-on-america.html.

"Rocket Principles," NASA, http://exploration.grc.nasa.gov/education/rocket/TRCRocket/rocket_principles.html.

Paul E. Ceruzzi, *Beyond the Limits: Flight Enters the Computer Age* (Cambridge, MA: MIT Press: 1989), 94–105.

"Syncom 3," NASA, http://nssdc.gsfc.nasa.gov/nmc/spacecraftDisplay.do?id=1964-047A.

"History of the NOAA Satellite Program," National Oceanic and Atmospheric Administration (updated June 2011), http://www.osd.noaa.gov/download/JRS012504-GD.pdf.

Rick W. Sturdevant, "NAVSTAR, the Global Positioning System: A Sampling of Its Military, Civil, and Commercial Impact," in *Societal Impact of Spaceflight,* ed. Steven J. Dick and Roger D. Launius (Washington, DC: National Aeronautics and Space Administration, 2009).

Up, Up, and Away

NAE Awards, Draper Prize 1991, http://www.nae.edu/Projects/Awards/DraperPrize/DraperWinners/page1991/55077.aspx.

"Origins of the Finite Element Method," Center for Aerospace Structures, University of Colorado Boulder, http://www.colorado.edu/engineering/cas/courses.d/IFEM.d/IFEM.AppO.d/IFEM.AppO.pdf.

Ray W. Clough and Edward L. Wilson, "Early Finite Element Research at Berkeley," Fifth U.S. National Conference on Computational Mechanics, August 4–6, 1999, http://www.ce.memphis.edu/7111/notes/class_notes/papers/fe-history.pdf.

"707 Family," Boeing, http://www.boeing.com/boeing/commercial/707family/index.page.

"Boeing 727-223," Museum of Flight, http://www.museumofflight.org/aircraft/boeing-727-223.

"Table 1-40: U.S. Passenger Miles (Millions)," Bureau of Transportation Statistics, U.S. Department of Transportation, http://www.rita.dot.gov/bts/sites/rita.dot.gov.bts/files/publications/national_transportation_statistics/html/table_01_40.html.

The World at Home: The Environmental Challenge

"Paul Muller–Biographical," Nobel Media AB 2014, http://www.nobelprize.org/nobel_prizes/medicine/laureates/1948/muller-bio.html.

"How Important Was Rachel Carson's *Silent Spring* in the Recovery of Bald Eagles and Other Bird Species?" *Scientific American,* August 21, 2012, http://www.scientificamerican.com/article/rachel-carson-silent-spring-1972-ddt-ban-birds-thrive/.

"The Pine River Statement: Human Health Consequences of DDT Use," Environmental Health Perspectives (2009), http://www.ncbi.nlm.nih.gov/pmc/articles/PMC2737010/.

"DDT–A Brief History and Status," U.S. Environmental Protection Agency, http://www.epa.gov/pesticides/factsheets/chemicals/ddt-brief-history-status.htm.

"Clean Air Act, United States," The Encyclopedia of Earth, http://www.eoearth.org/view/article/151129/.

Sharon Bertsch McGrayne, *Prometheans in the Lab: Chemistry and the Making of the Modern World* (New York: McGraw-Hill, 2001), 79–105.

George R. Tilton, *Clair Cameron Patterson, 1922–1995, A Biographical Memoir* (Washington, DC: National Academy of Sciences and National Academy Press, 1998), http://www.nasonline.org/publications/biographical-memoirs/memoir-pdfs/patterson-clair-c.pdf.

Patterson, Clair. Interview by Shirley K. Cohen. Pasadena, California, March 5, 6 and 9, 1995. Oral History Project, California Institute of Technology Archives, http://oralhistories.library.caltech.edu/32/1/OH_Patterson.pdf.

"1995 Tyler Laureate Clair C. Patterson," Tyler Prize for Environmental Achievement, http://www.usc.edu/dept/LAS/tylerprize/laureates/tyler1995.html.

"Lead Air Quality Standards," U.S. Environmental Protection Agency, http://www.epa.gov/oar/lead/standards.html.

"Leaded Gas Phaseout, Air Quality Fact Sheet, June 1995," U.S. Environmental Protection Agency, http://yosemite.epa.gov/R10/airpage.nsf/webpage/Leaded+Gas+Phaseout.

"CPSC Announces Final Ban on Lead-containing Paint," U.S. Consumer Product Safety Commission (September 2, 1977), http://www.cpsc.gov/en/Recalls/1977/CPSC-Announces-Final-Ban-On-Lead-Containing-Paint/.

Michael Rotman, "Cuyahoga River Fire," *Cleveland Historical,* http://clevelandhistorical.org/items/show/63#.Up5c7CfB-So.

Michael Scott, "Cuyahoga River Fire 40 Years Ago Ignited an Ongoing Campaign," *The Plain Dealer* (June 22, 2009), http://www.cleveland.com/science/index.ssf/2009/06/cuyahoga_river_fire_40_years_a.html.

"America's Sewage System and the Price of Optimism," *Time*, August 1, 1969, http://content.time.com/time/magazine/article/0,9171,901182,00.html.

"Clean Water Act, United States," The Encyclopedia of Earth, http://www.eoearth.org/view/article/51cbed3c7896bb431f690bc0/.

"Earth Day and EPA History," U.S. Environmental Protection Agency, http://www.epa.gov/earthday/history.htm.

To Your Health: The Engineering of Medical Imaging and Therapies

"General Nuclear Medicine," Radiological Society of North America, http://www.radiologyinfo.org/en/info.cfm?pg=gennuclear#part_five.

Gordon L. Brownell, "A History of Positron Imaging," October 15, 1999, http://neurosurgery.mgh.harvard.edu/docs/PEThistory.pdf.

"Hal Anger: Nuclear Medicine's Quiet Genius," *Journal of Nuclear Medicine* (vol. 44, no. 11 November 2003), http://interactive.snm.org/docs/hist_corner_11_03.pdf.

Constable and Somerville, *A Century of Innovation,* 158–161, 166–67.

"Tests and Procedures: MRI," Mayo Clinic, http://www.mayoclinic.org/tests-procedures/mri/basics/definition/prc-20012903.

"What is a PET Scan?" *Medical News Today,* http://www.medicalnewstoday.com/articles/154877.php.

Constable and Somerville, *A Century of Innovation,* 102–09.

John Adam, "Making Hearts Beat," *Innovative Lives,* http://invention.smithsonian.org/centerpieces/ilives/lecture09.html.

W.J. Kolff, "Artificial Organs, 40 Years and Beyond," http://www.stanford.edu/dept/HPS/transplant/html/kolff.html.

"Willem Kolff, Doctor Who Invented Kidney and Heart Machines, Dies at 97," *New York Times,* February 12, 2009, http://www.nytimes.com/2009/02/13/health/13kolff.html?pagewanted=all.

"Robert Jarvik on the Jarvik 7," http://www.jarvikheart.com/basic.asp?id=69.

1989: The Shrinking Globe

Introduction

Jarle Simensen, "Democracy and Globalization: 1989 and 'The Third Wave,'" *Journal of World History* 10, no. 2 (Fall 1999), 391–412.

"Fall of the Berlin Wall and Reunification 1989–1990," http://www.dw.de/fall-of-the-berlin-wall-and-reunification-1989-1990/a-16762692.

"Globalization and the Collapse of the Soviet Union," http://connection.ebscohost.com/c/articles/83534248/globalization-collapse-soviet-union.

Engineering and Human Welfare: Proceedings of the 25th Annual Meeting of the National Academy of Engineering, ed. Hedy E. Sladovich (Washington, DC: National Academy of Engineering, 1990), 10–11.

Building the Digital Highway

Ceruzzi, *Computing, 121–32.*

"A Brief History of the Internet," Internet Society, http://www.internetsociety.org/internet/what-internet/history-internet/brief-history-internet.

"Internet History–1980s," Computer History Museum, http://www.computerhistory.org/internet_history/internet_history_80s.html.

"The Launch of NSFNET," *America's Investment in the Future,* http://www.nsf.gov/about/history/nsf0050/internet/launch.htm.

"Networking: Vision and Packet-Switching 1959–1968, Intergalactic Vision to Arpanet," in James Pelkey, *Entrepreneurial Capitalism and Innovation: A History of Computer Communications 1968–1988 (2007),* http://www.historyofcomputercommunications.info/Book/2/2.1-IntergalacticNetwork_1962-1964.html.

"Memorandum For: Members and Affiliates of the Intergalactic Computer Network," http://worrydream.com/refs/Licklider-IntergalacticNetwork.pdf.

Constable and Somerville, *A Century of Innovation,* 146–53.

Dr. Lawrence G. Roberts, "Internet Chronology 1960–2001," http://www.packet.cc/internet.html.

TCP (Transmission Control Protocol)," *TechTarget,* http://searchnetworking.techtarget.com/definition/TCP.

"Introduction to TCP/IP," http://www.yale.edu/pclt/COMM/TCPIP.HTM.

"IP Address (Internet Protocol Address," http://searchunifiedcommunications.techtarget.com/definition/Internet-Protocol.

"A Brief History of the NSF and the Internet- Fact Sheet," August 13, 2003, http://www.nsf.gov/news/news_summ.jsp?cntn_id=103050.

Ray Tomlinson, "The First Network Email," http://openmap.bbn.com/-tomlinso/ray/firstemailframe.html.

"The History of Email," http://www.nethistory.info/History%20of%20the%20Internet/email.html.

"Batch processing," *Startup,* New Mexico Museum of Natural History and Sciences, http://startup.nmnaturalhistory.org/gallery/story.php?ii=39.

"Time-Sharing System Evolution," http://en.wikipedia.org/wiki/Time-sharing_system_evolution.

"Timesharing," *Startup,* New Mexico Museum of Natural History and Sciences, http://startup.nmnaturalhistory.org/gallery/story.php?ii=43&sid=3.

Birth of the World Wide Web

"Lunch with the FT: Tim Berners-Lee," September 7, 2012, http://www.ft.com/intl/cms/s/2/b022ff6c-f673-11e1-9fff-00144feabdc0.html#axzz2wr33vI9P.

"Where the Web Was Born," http://home.web.cern.ch/about/topics/birth-web/where-web-was-born.

"Berners-Lee: World Finally Realizes Web Belongs to No One," *Wired, June 6, 2012,* http://www.wired.com/wiredenterprise/2012/06/sir-tim-berners-lee/.

"History of the Web," World Wide Web Foundation, http://www.webfoundation.org/vision/history-of-the-web/.

"Information Management: A Proposal," Tim Berners-Lee, CERN, March 1989, May 1990, http://www.w3.org/History/1989/proposal.html.

"The World Wide Web Browser," http://www.w3.org/People/Berners-Lee/WorldWideWeb.html.

"Marc Andreessen, Internet Evangelist," *Entrepreneur, October 10, 2008,* http://www.entrepreneur.com/article/197600. (Marc Andreessen)

"Amazon Talks Strategy, International Growth," December 10, 2008, http://blog.seattlepi.com/amazon/2008/12/10/amazon-talks-strategy-international-growth/.

"History and Timeline of the ISS," Center for the Advancement of Science in Space (CASIS), http://www.iss-casis.org/About/ISSTimeline.aspx.

"Space Shuttle Program: Spanning 30 Years of Discovery," NASA, http://www.nasa.gov/mission_pages/shuttle/main/index.html.

"6 Coolest Space Shuttle Science Experiments," *Space.com* July 1, 2011, http://www.space.com/12150-6-coolest-space-shuttle-science-experiments.html.

Hubble Space Telescope, Science, NASA, http://www.nasa.gov/mission_pages/hubble/science/index.html#.U9YG5oBdUY2.

Clean and Efficient Energy

"OPEC," in Chapter 5 of *Modern Industrial Organization,* 4th ed., Dennis W. Carlton and Jeffrey M. Perloff (Prentiss Hall, 2004), http://wps.aw.com/aw_carltonper_modernio_4/21/5566/1424964.cw/content/index.html.

"Carl D. Keith, a Father of the Catalytic Converter, Dies at 88," *New York Times, November 15, 2008,* http://www.nytimes.com/2008/11/15/us/15keith.html.

"Historical Evolution of Powered Vehicles and Catalytic Converter," National Science Teachers Association, http://dev.nsta.org/evwebs/3368/History/History.htm.

"Driving to 54.5 MPG: The History of Fuel Economy," Clean Energy Project, Pew Project on National Security, Energy, and Climate (April 20, 2011), http://www.pewtrusts.org/en/research-and-analysis/fact-sheets/2011/04/20/driving-to-545-mpg-the-history-of-fuel-economy.

"Milestones in Mobile Source Air Pollution Control and Regulation," U.S. Environmental Protection Agency, Office of Transportation and Air Quality, http://www.epa.gov/otaq/consumer/milestones.htm.

"Catalytic Converters," *UCDavis ChemWiki,* http://chemwiki.ucdavis.edu/Physical_Chemistry/Kinetics/Case_Studies/Catalytic_Converters.

"Top 10 Improvements in Engine Design," http://auto.howstuffworks.com/10-improvements-in-engine-design.htm.

"Motor Vehicle Traffic Fatalities, 1900–2008," http://www.fhwa.dot.gov/policyinformation/statistics/2011/pdf/fi200.pdf.

"Air Quality Trends," U.S. Environmental Protection Agency, http://www.epa.gov/airtrends/aqtrends.html.

"Cleaning Up Coal: Knocking the NOx Out of Coal," U.S. Department of Energy, http://www.fossil.energy.gov/education/energylessons/coal/coal_cct3.html.

"Climate Change and Greenhouse Gases," in *Changing by Degrees: Steps to Reduce Greenhouse Gases, U.S. Office of Technology Assessment, 1991,* http://www.ciesin.org/docs/003-078/003-078.html.

"Overview of Greenhouse Gases," U.S. Environmental Protection Agency, http://www.epa.gov/climatechange/ghgemissions/gases/co2.html.

"Concentrating Solar Power: Energy from Mirrors," National Renewable Energy Laboratory, March 2001, http://www.nrel.gov/docs/fy01osti/28751.pdf.

"Concentrating Solar Power (CSP) Technologies," Solar Energy Development Programmatic EIS, http://solareis.anl.gov/guide/solar/csp/.

"Huge Thermal Plant Near Primm Opens as Solar Industry Grows," *Las Vegas Sun* February 13, 2014, http://www.lasvegassun.com/news/2014/feb/13/huge-thermal-plant-opens-solar-industry-grows.

"Interesting Wind Energy Facts," Wind Energy Foundation http://www.windenergyfoundation.org/interesting-wind-energy-facts.

"WindPACT," National Renewable Energy Laboratory, Wind Research, http://www.nrel.gov/wind/windpact.html.

"The History of Nuclear Energy," U.S. Department of Energy, Office of Nuclear Energy, Science and Technology, http://energy.gov/sites/prod/files/The%20History%20of%20Nuclear%20Energy_0.pdf.

"U.S. Nuclear Generating Statistics," Nuclear Energy Institute (updated April 2014), http://www.nei.org/Knowledge-Center/Nuclear-Statistics/US-Nuclear-Power-Plants/US-Nuclear-Generating-Statistics.

Building for Safety

"Historic Earthquakes," U.S. Geological Service, 1993, http://earthquake.usgs.gov/earthquakes/states/events/1989_10_18.php.

"San Francisco Passes Landmark Earthquake Retrofit Law," *SPUR*, April 24, 2013, http://www.spur.org/blog/2013-04-24/san-francisco-passes-landmark-earthquake-retrofit-law.

"Base Isolation: Origins and Development," National Information Service for Earthquake Engineering, http://nisee.berkeley.edu/lessons/kelly.html.

"William Henry Robinson, Citation on the Award of Life Membership, April 2008," New Zealand Society for Earthquake Engineering Inc., http://www.nzsee.org.nz/membership/distinguished-members/william-henry-robinson/.

Robert K. Reitherman, *Earthquakes and Engineers: An International History* (Reston, VA: American Society of Civil Engineers, 2012), 380–83.

"Seismic Design Hailed for Averting Hospital Damage," *Los Angeles Times,* March 24, 1994, http://articles.latimes.com/1994-03-24/news/mn-37881_1_base-isolation.

"City Hall Getting Latest Quake Upgrade," *Los Angeles Times, March 22, 1995,* http://articles.latimes.com/1995-03-22/local/me-45619_1_base-isolation.

"A Tested, Inexpensive Way to Protect Buildings from Earthquakes," *NOVA Next,* http://www.pbs.org/wgbh/nova/next/tech/rubber-bearings-seismic-protection/.

"Largest High-Rise Seismic Sensor Network in CA installed in One Rincon Hill," *The San Francisco Appeal,* May 30, 2012, http://sfappeal.com/2012/05/largest-high-rise-seismic-sensor-network-in-ca-installed-in-one-rincon-hill/.

"Seismic Analysis Using Finite Elements," International Association of the Engineering Modeling, Analysis and Simulation Community (NAFEMS), http://www.nafems.org/events/nafems/2008/seismic_dec/.

"UC Berkeley Shaking Table," Pacific Earthquake Engineering Research Center, http://peer.berkeley.edu/laboratories/shaking_table.html.

"Full-on testing at full-scale earthquake testing facility," *ElectronicsWeekly.com,* http://www.electronicsweekly.com/made-by-monkeys/test/full-on-testing-at-full-scale-2012-06/.

"Progress Toward a Safer Future Since the 1989 Loma Prieta Earthquake," U.S. Geological Service (Fact Sheet 151 created November 3,1999, updated May 6, 2005), http://pubs.usgs.gov/fs/1999/fs151-99/.

Knowing Ourselves: Biomedical Engineering

"Paul Berg Creates First Recombinant DNA Molecules," Genetics and Genomics Timeline, *Genome News Network,* http://www.genomenewsnetwork.org/resources/timeline/1972_Berg.php.

"Milestones in Biotechnology," *Genetic Engineering & Biotechnology News,* http://www.genengnews.com/timeline/#null.

"A Brief History of Genetic Modification," *gmeducation.org,* http://www.gmeducation.org/faqs/p149248-a-brief-history-of-genetic-modification.html.

"Recombinant DNA: Example Using Insulin," *Explore More,* http://www.iptv.org/exploremore/ge/what/insulin.cfm.

"Genomes," *Genome News Network,* http://www.genomenewsnetwork.org/resources/whats_a_genome/Chp1_1_1.shtml#genome1.

http://www.genome.gov/Pages/Education/AllAbouttheHumanGenomeProject/GuidetoYourGenome07_vs2.pdf.

"An Introduction to DNA and Chromosomes," Huntington's Outreach Project for Education at Stanford (HOPES), https://www.stanford.edu/group/hopes/cgi-bin/wordpress/2011/02/an-introduction-to-dna-and-chromosomes-text-and-audio/#what-are-complementary-strands-making-the-double-helix.

"The Human Genome Project Completion: Frequently Asked Questions," National Human Genome Research Institute, http://www.genome.gov/11006943.

"The Human Genome Project," Genetics Home Reference, http://ghr.nlm.nih.gov/handbook/hgp?show=all.

Human Genome Project Information Archive, 1990–2003, http://web.ornl.gov/sci/techresources/Human_Genome/index.shtml.

"A History of the Human Genome Project," *Science,* 291, no. 5507 (February 16, 2001), 1195, http://www.sciencemag.org/content/291/5507/1195.

"The Human Genetic Code—The Human Genome Project and Beyond," Center for Genetics Education, http://www.genetics.edu.au/Publications-and-Resources/Genetics-Fact-Sheets/FactSheet24.

"Scientists Complete the First Draft of Human DNA Blueprint," Human Genome Project, *The Guardian.com,* June 26, 2000, http://www.theguardian.com/science/human-genome-project.

"Machines and the Human Genome Project," Lawrence Berkley National Laboratory, Spring 1992, http://www.lbl.gov/Science-Articles/Archive/human-genome-mapping-sequencing.html.

"Computers and the Human Genome Project," http://www-cs-faculty.stanford.edu/-eroberts/courses/soco/projects/2000-01/computers-and-the-hgp/.

"Sequencing Human Genome: The Contributions of Francis Collins and Craig Venter," *Nature Education* 1(1):133, http://www.nature.com/scitable/topicpage/sequencing-human-genome-the-contributions-of-francis-686.

J. Craig Venter Biography, http://www.jcvi.org/cms/about/bios/jcventer/.

"International Human Genome Sequencing Consortium Describes Finished Human Genome Sequence," National Human Genome Research Institute, October 20, 2004, http://www.genome.gov/12513430.

"Winner: The Revolution Will Be Prosthetized," *Spectrum, IEEE, January 1, 2009,* http://spectrum.ieee.org/robotics/medical-robots/winner-the-revolution-will-be-prosthetized.

"Towards an EMG Controlled Prosthetic Hand Using a 3-D Electromagnetic Positioning System," 2007, http://www.articles-abstracts.com/article.php?id=498706&author=Yu_Su_Fisher_M_H_Wolczowski_A_Bell_G_D_Burn_D_J_Gao_R_X&title=Towards_an_EMG_Controlled_Prosthetic_Hand_Using_a_3_D_Electromagnetic_Positioning_System.

"Tissue Engineering," *Nature Biotechnology* 18, IT56–IT58 (2000) http://www.nature.com/nbt/journal/v18/n10s/full/nbt1000_IT56.html.

"Tissue Engineering: How to Build a Heart," *Nature,* July 3, 2013, http://www.nature.com/news/tissue-engineering-how-to-build-a-heart-1.13327.

2014: A Healthier, Cleaner, More Connected World

Introduction

ICT Facts and Figures 2014, http://www.itu.int/en/ITU-D/Statistics/Pages/stat/default.aspx.

Tiny Powerhouses, Global Reach

"60 Years of the Transistor: 1947–2007," http://www.intel.com/cn/technology/timeline.pdf.

"Ultrafast Deep UV Lithography with Excimer Lasers," *Electron Device Letters*, IEEE 3, no. 3 (March 1982), http://ieeexplore.ieee.org/xpl/articleDetails.jsp?reload=true&arnumber=1482581.

"Magnetic Field of Dreams," *Businessweek*, April 17, 1994, http://www.businessweek.com/stories/1994-04-17/magnetic-field-of-dreams.

"Creators of Lithium-Ion Battery and Pioneers of Innovative Education Model Win Engineering's Highest Honors," National Academy of Engineering, January 2014, http://www.nae.edu/Projects/MediaRoom/20095/69135/106261.aspx.

"What a Cell Phone Can Do for a Nigerian Farmer: Quail," *Radio Netherlands Worldwide*, February 12, 2013, http://www.rnw.nl/africa/article/what-a-cell-phone-can-do-a-nigerian-farmer-quail.

"Fast Internet Is Chattanooga's New Locomotive," *New York Times*, February 4, 2014, http://www.nytimes.com/2014/02/04/technology/fast-internet-service-speeds-business-development-in-chattanooga.html?ref=todayspaper&_r=0.

"New Study Quantifies Use of Social Media in Arab Spring," University of Washington, September 2011, http://www.washington.edu/news/2011/09/12/new-study-quantifies-use-of-social-media-in-arab-spring/.

A Cleaner World

"Acid Rain Imperils Adirondacks Fish," *New York Times*, July 7, 1988, http://www.nytimes.com/1988/07/07/nyregion/acid-rain-imperils-adirondacks-fish.html.

"How Much Energy Does a Person Use in a Year?" U.S. Energy Information Administration, http://www.eia.gov/tools/faqs/faq.cfm?id=85&t=1.

"Energy Intensity," U.S. Energy Information Administration, December 2013, http://www.eia.gov/forecasts/aeo/er/early_intensity.cfm.

"Progress Cleaning the Air and Improving People's Health," U.S. Environmental Protection Agency, http://www.epa.gov/air/caa/progress.html#clean.

"Low-Temperature Combustion Enables Cleaner, More Efficient Engines," Sandia National Laboratories, August 2013, https://share.sandia.gov/news/resources/news_releases/ltc_diesel/#.UurHvCiChLU.

Tier 3 Motor Vehicle Emission and Fuel Standards, Congressional Research Service 7-5700 R43497, April 28, 2014, http://nationalaglawcenter.org/wp-content/uploads/asset/crs/R43497.pdf.

"New-Car Average Fuel Economy Rose to 24.8 MPG in 2013," *Green Car Reports*, January 9, 2014, http://www.greencarreports.com/news/1089565_new-car-average-fuel-economy-rose-to-24-8-mpg-in-2013.

"South Coast Air Basin Smog Trend," South Coast Air Quality Management District, http://www.aqmd.gov/docs/default-source/air-quality/historical-data-by-year/ozone-trend-chart-from-historical-data-by-year.pdf.

"2014 Most and Least Efficient Vehicles," U.S. Department of Energy, http://www.fueleconomy.gov/feg/best-worst.shtml.

"Cross Country Rally: By the Numbers," Tesla Motors, February 2014, http://www.teslamotors.com/blog/cross-country-rally-numbers.

"Building Energy Efficiency Program," California Energy Commission, http://www.energy.ca.gov/title24/.

"California Energy Consumption," U.S. Department of Energy, http://apps1.eere.energy.gov/states/consumption.cfm/state=CA.

"How a Company Found a Business in Measuring Leaking Methane," *ClimateWire*, August 2, 2013, http://www.eenews.net/stories/1059985507.

"Professor Ferdinand Porsche Created the First Functional Hybrid Car," Porsche Press Release, April 20, 2011, http://press.porsche.com/news/release.php?id=642.

"Cars of Futures Past," *Hemmings Daily,* August 15, 2013, http://blog.hemmings.com/index.php/2013/08/15/cars-of-futures-past-1901-lohner-porsche-semper-vivus-and-mixte/.

"From Model T to Prius: 13 Big Moments in Fuel Efficiency History," *Christian Science Monitor,* March 7, 2012, http://www.csmonitor.com/Innovation/2012/0307/From-Model-T-to-Prius-13-big-moments-in-fuel-efficiency-history.

"Toyota Prius Chronological History," http://www.toyoland.com/prius/chronology.html.

"Light Duty Electric Drive Vehicles Monthly Sales Updates," Argonne National Laboratory, Transportation R and D Center, http://www.transportation.anl.gov/technology_analysis/edrive_vehicle_monthly_sales.html.

"Hybrid and Plug-In Electric Vehicles," U.S. Department of Energy, October 2011, http://greaterindiana.brandrenew.net/wp-content/uploads/2012/01/Hybrid-and-EV.pdf.

"Gas Prices Hit Highest Average Ever in 2012," CNN Money, December 31, 2012, http://money.cnn.com/2012/12/31/news/economy/gas-prices/.

"Energy Department Releases Updated eGallon Prices as Electric Vehicle Sales Double," *Energy.gov,* July 19, 2013, http://energy.gov/articles/energy-department-releases-updated-egallon-prices-electric-vehicle-sales-double.

"Dan Abraham," Argonne National Laboratory, February 5, 2012, http://www.anl.gov/articles/dan-abraham.

An Unexpected Change in the Energy Landscape: The Shale Gas Boom

Gregory Zuckerman, *The Frackers: The Outrageous Inside Story of the New Billionaire Wildcatters* (New York: Portfolio/Penguin, 2013).

"U.S. Expected to Be Largest Producer of Petroleum and Natural Gas Hydrocarbons in 2013," U.S. Energy Information Administration, October 4, 2013, http://www.eia.gov/todayinenergy/detail.cfm?id=13251.

Coal Data Browser, U.S. Energy Information Administration, http://www.eia.gov/coal/data.cfm#consumption.

"U.S. Greenhouse Gas Emissions Fall 10 Percent Since 2005: EPA," *Reuters*, April 15, 2014, http://www.reuters.com/article/2014/04/15/us-usa-climate-carbon-idUSBREA3E1XD20140415.

"National Climate Assessment," U.S. Global Change Research Program, 2014, http://nca2014.globalchange.gov.

"Gas Pains," *Conservation*, June 2013, http:conservationmagazine.org/2013/06/gas-pains/.

Renewable Energy

"The 10 Biggest Turbines in the World," *Wind Power Monthly,* http://www.windpowermonthly.com/10-biggest-turbines.

"Wind Topped All New U.S. Generation Capacity in 2012, Ahead of Gas and Coal," *Greentech Media*, February 7, 2013, http://www.greentechmedia.com/articles/read/Wind-Topped-All-New-U.S.-Generation-Capacity-in-2012-Ahead-of-Gas-and-Coal.

"Are We on the Cusp of a Solar Energy Boom?" *The Week*, May 21, 2013, http://theweek.com/article/index/244437/are-we-on-the-cusp-of-a-solar-energy-boom.

"Photovoltaic (PV) Pricing Trends: Historical, Recent, and Near-Term Projections," U.S. Department of Energy, November 2012, http://www.nrel.gov/docs/fy13osti/56776.pdf.

Federal Energy Regulatory Commission, "Energy Infrastructure Update," December 2013, http://www.ferc.gov/legal/staff-reports/2013/dec-energy-infrastructure.pdf.

"Short-Term Energy Outlook," U.S. Energy Information Administration, July 8, 2014, http://www.eia.gov/forecasts/steo/report/electricity.cfm.

"How Much Gasoline Does the United States Consume?" U.S. Energy Information Administration, http://www.eia.gov/tools/faqs/faq.cfm?id=23&t=10.

"Denmark, Portugal, and Spain Leading the World in Wind Power," Earth Policy Institute, May 2014, http://www.earth-policy.org/data_highlights/2014/highlights46.

A Healthier World

Paul Citron, retired, vice president, Technology Policy and Academic Relations, Medtronic, interviewed by John A. Carey, December 2013.

"How Medtronic Made the World's Smallest Pacemaker," *Cardiovascular*, December 2013, http://www.qmed.com/mpmn/medtechpulse/how-medtronic-made-worlds-smallest-pacemaker.

"Outpatient Glycemic Control with a Bionic Pancreas in Type 1 Diabetes," *New England Journal of Medicine*, June 15, 2014, http://www.nejm.org/doi/full/10.1056/NEJMoa1314474#t=article.

"Life Expectancy: Data by Country," World Health Organization, http://apps.who.int/gho/data/node.main.688?lang=en.

"Mortality Rate," The World Bank, http://data.worldbank.org/indicator/SH.DYN.MORT.

"Declining Mortality from Smoking in the United States," *Nicotine & Tobacco Research,* July 2007, http://www.ncbi.nlm.nih.gov/pubmed/17577808.

"Fatality Analysis Reporting System," National Highway Safety Traffic Administration, http://www-fars.nhtsa.dot.gov/Main/index.aspx.

"Sick Water? The Central Role of Waste-Water Management in Sustainable Development." United Nations Environment Programme, UN-HABITAT, 2010, http://www.unep.org/pdf/SickWater_screen.pdf.

"How the World's Oceans Could Be Running Out of Fish," BBC, September 2012, http://www.bbc.com/future/story/20120920-are-we-running-out-of-fish.

Looking to the Future

Introduction

"Bad Predictions," *Things People Said,* http://www.rinkworks.com/said/predictions.shtml.

"Predictions of the Future from the Past," *PandoDaily,* January 2014, http://pando.com/2014/01/03/predictions-of-the-future-from-the-past/.

"Visit to the World's Fair of 2014," *New York Times*, August 1964, http://www.nytimes.com/books/97/03/23/lifetimes/asi-v-fair.html.

Brought to You by Engineering

"Grand Challenges for Engineering," National Academy of Engineering, 2008, http://www.engineeringchallenges.org/cms/challenges.aspx.

"Alan Kay: Educator and computing pioneer," TED Speaker, http://www.ted.com/speakers/alan_kay.

New Materials, New Possibilities

"Graphene—The Perfect Atomic Lattice," The Nobel Prize in Physics 2010, Information for the Public, http://www.nobelprize.org/nobel_prizes/physics/laureates/2010/popular-physicsprize2010.pdf.

"Bend It, Charge It, Dunk It: Graphene, the Material of Tomorrow," *New York Times, April 2014,* http://bits.blogs.nytimes.com/2014/04/13/bend-it-charge-it-dunk-it-graphene-the-material-of-tomorrow/?_php=true&_type=blogs&_php=true&_type=blogs&emc=eta1&_r=1&.

"Stan Williams," HP Fellows, http://www.hpl.hp.com/about/honors/HPfellows/williams.html.

"Physics: 'Putting the Weirdness to Work,'" *Businessweek*, March 2004, http://www.businessweek.com/stories/2004-03-14/physics-putting-the-weirdness-to-work.

Dr. Venkatesh Narayanamurti, Benjamin Peirce Professor of Technology and Public Policy Harvard School of Engineering and Applied Sciences, interview by John A. Carey, February 2014.

A World of Embedded Intelligence

"Auto Correct: Has the Self-Driving Car At Last Arrived?" *The New Yorker*, November 2013, http://www.newyorker.com/reporting/2013/11/25/131125fa_fact_bilger?currentPage=all.

Dr. Leah H. Jamieson, John A. Edwardson Dean of Engineering and Ransburg Distinguished Professor of Electrical and Computer Engineering, Purdue University, interview by John A. Carey, May 2014.

"Ray Kurzweil: This is Your Future," CNN, December 2013, http://edition.cnn.com/2013/12/10/business/ray-kurzweil-future-of-human-life/.

Dr. W. Daniel Hillis, chairman and cofounder Applied Minds, LLC, interview with John A. Carey, February 2014.

Expanding Our Connections

Charles O. Holliday Jr., retired chairman of the board and CEO, E.I. du Pont de Nemours & Company, interview with John A. Carey, December 2013.

"Digital Life in 2025," Pew Research Center, March 2014, http://www.pewinternet.org/2014/03/11/digital-life-in-2025/.

"Supersize Me: GE Takes 3D Printing to Massive Gas Turbines," *GE Reports*, July 2013, http://www.gereports.com/post/74545205358/supersize-me-ge-takes-3d-printing-to- massive-gas.

Making Energy Sustainable

"Climate Change: Evidence & Causes," The Royal Society and the U.S. National Academy of Sciences, 2014, https://royalsociety.org/-/media/Royal_Society_Content/policy/projects/climate-evidence-causes/climate-change-evidence-causes.pdf.

"National Climate Assessment," http://nca2014.globalchange.gov.

"U.S. Quest for 5X Battery Boost Seeks 'Game-Changing' Tech," *Computerworld*, 2013, http://www.computerworld.com/s/article/9244074/U.S._quest_for_5X_battery_boost_seeks_game_changing_tech.

"Welcome to Open Source Aerogel," http://www.aerogel.org.

Dr. John C. Wall, vice president and chief technical officer, Cummins, interview with John A. Carey, February 2014.

Dr. Julia M. Phillips, vice president and chief technology officer, Sandia National Laboratories, interview with John A. Carey, January 2014.

"Solar Energy Conversion," Nocera Lab, http://nocera.harvard.edu/SolarEnergyConversion.

"Cheap Hydrogen from Sunlight and Water," *Technology Review*, 2013, http://www.technologyreview.com/news/521671/cheap-hydrogen-from-sunlight-and-water/.

"DOE Energy Innovation Hubs," U.S. Department of Energy, http://science.energy.gov/bes/research/doe-energy-innovation-hubs/.

Dr. Cherry Murray, dean of the Harvard School of Engineering and Applied Sciences, and John A. and Elizabeth S. Armstrong Professor of Engineering and Applied Sciences, interview with John A. Carey, May 2014.

"Global Coal Risk Assessment," World Resources Institute, 2012, http://pdf.wri.org/global_coal_risk_assessment.pdf.

Dr. Robert H. Socolow, professor of mechanical and aerospace engineering, Princeton Environmental Institute, Princeton University, interview with John A. Carey, February 2014.

Feeding the World's Billions

Dr. J. Craig Venter, chairman and president, J. Craig Venter Institute, interview with John A. Carey, February 2014.

Dr. Cherry Murray, May 2014.

A Healthier Future

Dr. Robert Samuel Lander, David H. Koch Institute Professor, Massachusetts Institute of Technology, interview with John A. Carey, December 2013.

"Karl Deisseroth, MD, PhD," Howard Hughes Medical Institute, http://www.hhmi.org/scientists/karl-deisseroth.

Dr. W. Daniel Hillis, February 2014.

Paul Citron, December 2013.

Dr. J. Craig Venter, February 2014.

Cities, Limits, and New Frontiers

"World Population Projected to Reach 9.6 Billion by 2050," United Nations Department of Economic and Social Affairs, 2013, https://www.un.org/en/development/desa/news/population/un-report-world-population-projected-to-reach-9-6-billion-by-2050.html.

"World Population Prospects," United Nations, 2013, http://esa.un.org/unpd/wpp/Documentation/pdf/WPP2012_HIGHLIGHTS.pdf.

"10 Projections for the Global Population in 2050," Pew Research Center, http://www.pewresearch.org/fact-tank/2014/02/03/10-projections-for-the-global-population-in-2050/.

"Planning the City of the Future," News Briefing, AAAS Annual Meeting, 2014, http://www.eurekalert.org/aaasnewsroom/2014/watchbriefing.php?page=26.

"Is There a Limit to How Tall Buildings Can Get?" *Atlantic CityLab*, 2012, http://www.citylab.com/design/2012/08/there-limit-how-tall-buildings-can-get/2963/.

Charles O. Holliday, Jr., December 2013.

"Livable Cities of the Future," National Academy of Engineering, National Academies Press, 2014. https://www.nae.edu/113563.aspx.

NYC Special Initiative for Rebuilding and Resiliency, http://www.nyc.gov/html/sirr/html/home/home.shtml.

Dr. Wanda M. Austin, president and CEO, The Aerospace Corporation, interview with John A. Carey, May 2014.

Dr. Robert H. Socolow, February 2014.

"Steven Pinker TED Speaker Personal Profile," TED, http://www.ted.com/speakers/steven_pinker.

"NASA's Kepler Discovers First Earth-Size Planet In The 'Habitable Zone' of Another Star," NASA, April 17, 2014, http://www.nasa.gov/ames/kepler/nasas-kepler-discovers-first-earth-size-planet-in-the-habitable-zone-of-another-star/#.U8_8VoBdXes.

"Kuiper Belt and Oort Cloud," NASA, Solar System Exploration, http://solarsystem.nasa.gov/planets/profile.cfm?Object=KBOs&Display=OverviewLong.

"The FOCAL Mission: To the Sun's Gravity Lens," *Centauri Dreams,* August 18, 2006, http://www.centauri-dreams.org/?p=785.

"What is Gravitational Lensing?" The Canada France Hawai'i Telescope Lensing Survey (CFHTLenS), http://www.cfhtlens.org/public/what-gravitational-lensing.

"The Seventy-Billion-Mile Telescope," http://www.newyorker.com/online/blogs/elements/2013/06/gravitational-lens-sun-space-exploration-maccone.html.

"The Solar Wind," NASA, http://solarscience.msfc.nasa.gov/SolarWind.shtml.

"Interstellar Journey: Shrinking the Probe," *Centauri Dreams,* July 14, 2014, http://www.centauri-dreams.org/?p=31049.

"Space Exploration for Everyone," http://pocketspacecraft.com/.

Dr. David Messerschmitt, Roger A. Strauch Professor Emeritus of Electrical Engineering and Computer Sciences, University of California, Berkeley, interview with Roberta Conlan, July 2014.

PHOTO CREDITS

Page 1, (top left) Victor R. Boswell, Jr./National Geographic/Getty Images, (top right) Toria/Shutterstock, (bottom left) diversepixel/Shutterstock, (bottom right) wellphoto/Shutterstock; page 2, Courtesy University of Maryland; page 4, Songquan Deng/Shutterstock; page 5, (left) © CORBIS, (right) Creative Commons Share-alike; page 6, (top left) Creative Commons Share-alike, (top right) Courtesy of the Computer History Museum, (bottom left) Elle1/Shutterstock; page 7, (top right) AP Photo/IBM, (middle left) Courtesy Texas Instruments, (middle right) Official U.S. Navy Photograph/NHHC, (bottom left) © Mark Richards/Courtesy of the Computer History Museum, (bottom right) Creative Commons Share-alike; page 8, (bottom left) Rauner Library, Dartmouth College, (middle) Courtesy of the Computer History Museum, (top right) Courtesy of the Computer History Museum, (bottom right) Courtesy of the Computer History Museum; page 9, PeterPhoto123/Shutterstock; page 10, (left) Dmitry Kalinovsky/Shutterstock, (right) Monkey Business Images/Shutterstock, (background) X=etra/Shtterstock; page 11, (left) NASA, (right) NASA; page 12, (top) NASA, (bottom)NOAA; page 13, (top) Ivan Sholakov/Shutterstock, (bottom left) USAF, (bottom right) Imperial War Museum, London; page 14, tumblr; page 15, (top) Landov Media, (bottom) AP Photo/Mark Duncan; page 16, (top) Photonics Media/Courtesy of Lawrence Berkeley National Laboratory, (bottom) © Bettman/CORBIS; page 17, (top) Reproduced with permission of Medtronic, Inc., (bottom left) University of Buffalo, (bottom right) © Diego Goldberg/Sygma/CORBIS ; page 18, SSokolov/Shutterstock; page 19, (top) Photo by Ben Schneiderman/Courtesy of the Computer History Museum, (bottom) Courtesy of the Computer History Museum; page 20, (left) Raytheon BBN Technologies, (right) Copyright NCSA/UIUC, image by Donna Cox and Robert Patterson; page 21, Courtesy of the Computer History Museum; page 22, (left) CERN, (right) CERN; page 23, (bottom left) NASA, (top right) Library of Congress; page 24, (top left) National Science and Technology Medals Foundation, (top right) © Science Photo Library/Science Photo Library/Corbis, (bottom) Ken Schulze/Shutterstock; page 25, National Renewable Energy Laboratory; page 26, (left) © Grant Smith/CORBIS, (right) Creative Commons Share-alike; page 27, © Karen Kasmauski/CORBIS; page 28, (top left) National Human Genome Research Institute, (top right) J. Craig Venter Institute, (bottom left) Organogenesis, Inc., (bottom right) Tony Taylor Stock/Shutterstock; page 29, PHOTOCREO Michal Bednarek/Shutterstock; page 30, (left) Tim McNerney/Intel Museum/Creative Commons Attribution-Noncommercial-Share Alike 3.0 License, (right) ChrisVanLennepPhoto/Shutterstock; page 31, Creative Commons Share-alike; page 32, (bottom left) Wikipedia/Creative Commons Share-alike, (bottom middle) Oakridge National Laboratory, (bottom right) Bloom Design/Shutterstock, (top left) Photo by Andrey Shcherbina, Woods Hole Oceanographic Institution, (top right) Denys Prykhodov/Shutterstock; page 33, (left top) Straight 8 Photography/Shutterstock, (left bottom) Andrey_Popov/Shutterstock, (right) Nolte Lourens/Shutterstock; page 34, (top left) MNStudio/Shutterstock, (top right) Dennis Donohue/Shutterstock, (bottom) Photo by Dino Vournas/Sandia National Laboratories; page 35, (top) Tesla Motors, (bottom) Tesla Motors; page 36, (top left) Creative Commons Share-alike, (top right) Courtesy of Toyota, (bottom) Wikipedia/Creative Commons Share-alike; page 37, (left) Creative Commons Share-alike, (right) U.S. Energy Information Administration; page 38, U.S. Energy Information Administration; page 39, Sandia Labs/Flickr Photo album; page 40, (top) Creative Commons Share-alike, (bottom) Sonquan Deng/Shutterstock; page 41, (left)Monkey Business Images/Shutterstock, (middle) Picsfive/Shutterstock, (right) Reproduced with permission of Medtronic, Inc.; page 42, (left) JHU/APL, (right) Maksim Vivtsaruk/Shutterstock; page 43, (top left) Sura Nualpradid/Shutterstock, (top right) Maksim Vivtsaruk/Shutterstock, (bottom) otiki/Shutterstock; page 44, Sarah Holmlund/Shutterstock; page 46, (left top) Wikipedia/Creative Commons Attribution 3.0 License, (left bottom) AP Photo/Jon Super, (right) HP Labs; page 47, (top left) Slavoljub Pantelic/Shutterstock, (top right) DARPA/US Government, (middle right) Photo by PHC(NAO) Chris Desmond COMNAVCRUITCOM, (bottom) Wikipedia/Creative Commons Attribution 3.0 License; page 48, Babich Alexander/Shutterstock; page 49, (top left) KT Crabb Photography, (middle) KT Crabb Photography, (top right) Peshkova/Shutterstock, (bottom left) Creative Commons Share-alike; page 50, (left) Sandy Huffaker for The New York Times/Redux, (right) Creative Commons Share-alike; page 51, (top) Vladimir Wrangel/Shutterstock, (bottom left) Monica Wisniewska/Shutterstock, (bottom right) Bork/Shutterstock; page 52, IndustryAndTravel/Shutterstock; page 53, (left) Sergey Nivens/Shutterstock, (right) shaziafolio/Shutterstock; page 54, leungchopan/Shutterstock; page 55, (top) Adrian Mann/Centauri Dreams, (bottom) NASA/Marshall Space Flight Center.